Alexandre Santos

Development of a self-guided quadcopter using GPS signals

Alexandre Santos

Development of a self-guided quadcopter using GPS signals

Assembly and configuration of the quadcopter (drone)

ScienciaScripts

Imprint

Cover image: www.ingimage.com

This book is a translation from the original published under ISBN 978-613-9-72420-8.

Publisher:
Sciencia Scripts
is a trademark of
Dodo Books Indian Ocean Ltd. and OmniScriptum S.R.L publishing group

120 High Road, East Finchley, London, N2 9ED, United Kingdom
Str. Armeneasca 28/1, office 1, Chisinau MD-2012, Republic of Moldova, Europe
Printed at: see last page
ISBN: 978-620-7-92787-6

ACKNOWLEDGEMENTS

Life is full of challenges. Challenges that we must try to overcome at all times, preferably alongside the people who are always present in our lives, helping and guiding us when we need it most. For this reason, I thank you:

To my family for their trust.

To my friends for their support and encouragement.

To the teachers for their patience and teaching.

I would especially like to thank my supervisor Prof Lucas Kriesel Sperotto, because whenever I needed him, he was ready to help me and answer all my questions, as well as his great theoretical teachings and his profound demonstration of knowledge in order to build this work and complete this challenge.

"Try *once, twice, three times and if possible try a fourth, a fifth and as many times as necessary. Just don't give up at the first attempt, persistence is the friend of achievement. If you want to get where most people don't, do what most people don't."*

(Bill Gates)

SUMMARY

This project relates to the concepts of control and automation, analogue and digital electronics, microcontrollers, embedded systems, mechatronics, robotics and radio frequency signals. In order to build a quadcopter capable of flying over windy areas and recording information, we will need all these sophistications. When the quadcopter takes off, its first function will be to stabilise itself so that there are no accidents in strong winds or that it doesn't interfere with filming or recording photos. To do this, we will use a controller board programmed in Arduino. After stabilising itself, the controller board will provide the function of guiding by radio frequency control or self-guiding by GPS route.

Keywords: Quadcopter. Drone. UAV, APM board. Mission Planner.

SUMMARY

INTRODUCTION

When people come across an Unmanned Aerial Vehicle (UAV), they start to pay attention to the manoeuvres during its flight. This attention arises spontaneously. This arouses curiosity about what it would be like to use a UAV. It was precisely this curiosity that motivated the research, as we saw the possibility of carrying out a study on how model aeroplanes work, especially helicopters and quadcopters, as well as their respective aerodynamics. In addition, the work could serve as a basis for future academics who want to develop their research in the areas of robotics or control and automation.

The aim was to design and build a low-cost quadcopter that could steer itself via georeferencing. This included mechanical and electronic projects, as well as the development of an embedded system and quadcopter control software. This project began with the construction of a quadcopter using the ArduPilot Mega (APM) controller board. The initial intention was based on projects that could contribute in some way to the development of rural and commercial areas by preserving and monitoring the environment.

To give veracity and basis to this work, we used theorists such as Studart with the work Quadricopter learning a little more about this marvellous machine, Vettorazzi describing the Global Positioning System, Vieira with QuadRotor aerial mobile platform, Medeiros reporting on the Development of an unmanned aerial vehicle for application in precision agriculture, among others.

The development of this project could serve as a basis for solving problems with monitoring large areas, as well as small areas that need more attention. Problems related to town planning, security and maintenance of areas that are difficult to access can be easily solved with secure monitoring.

By visualising the images generated by the model aircraft, the operator will have a full understanding of what is happening in real time, so they can make the right decision to solve the problem on their property. What's more, the cost will be much lower than if they used a helicopter or any other type of aircraft to supervise their property.

According to Keanee Carr (2013), UAVs have been developed for over 100 years, even before the First World War. Engineers had already been trying to find a way to build a remotely controlled aircraft. The achievement of developing a UAV became ever closer. As technology advanced, a fixed-wing model aeroplane appeared that could be controlled. This model aircraft began to be used in military projects.

The first UAVs with rotating wings were popularly known as drones. The name Drone came about because of the noise of their propellers, which sounded like the buzzing of a drone. The most widely used model is the quadcopter. This model has four propellers, one for each of its engines. According to Studart (2015), the aeromodelo was seen as a solution for vertical flights, as it had good stability, which can be seen in just a few minutes of flying.

With a wide choice of model aeroplanes for different purposes, the quadcopter stands out for its simple construction and cheap parts, as well as its agility. These advantages make it an excellent choice for aerial inspection services. When using a helicopter or other aircraft, the cost is higher.

The choice of a model aeroplane can be considered very versatile, as it can be controlled simply by a radio control or via a smartphone application, thus enabling various operations and manoeuvres. The model aircraft uses electronic systems and sensors to receive its commands, meaning it doesn't need a professional pilot to make it move and stabilise. However, it is necessary to have knowledge of the flight controller and all its operations.

The quadcopter has four rotors. These rotors generate the movement of the propellers, thus initiating its hovering. This type of aeroplane differs from helicopters and fixed-wing UAVs in that its engines have four propellers to generate propulsion. Helicopters, on the other hand, have a main propeller for sustainability and a propeller for stabilisation when the angular force is generated by the rotation of the main propeller, which causes it to remain stationary on its axis of rotation. Keeping control of a helicopter or model aeroplane is no easy task.

Quadcopters have a system with more than one propeller that cancels out angular forces. The quadcopter moves using a signal sent by radio frequency telemetry and received by software installed in the flight controller, which passes the information on to all the drone's components.

One of the best-known and most widely used controller boards in the UAV market is the ArduPilot Mega (APM), which controls all the UAV's functions and is the main component for controlling the model aeroplane, communicating with the user via the Mission Planner software. This board is the brain of the UAV and has several microcontrollers in its electronic system. The details of the board can be seen in Chapter 2, in which all its usefulness will be described.

This project began with the construction of a low-cost quadcopter using the APM controller board. The initial intention was based on projects that could contribute in some way to the development of rural and commercial areas by preserving and monitoring the environment. To give veracity and basis to this work, we used theorists such as Studart with his work Quadcopter, learning a little more about this marvellous machine, Vettorazzi describing the Global Positioning System, Vieira with the QuadRotor aerial mobile platform, among others. The work is organised as follows:

Chapter 1 refers to the historical context of UAVs, with the development and advancement of UAV technologies along with the standards that are used in Brazil. Chapter 2 details all the components used to assemble the Quadcopter. Chapter 3 presents the installation and configuration of the flight controller software, the step-by-step assembly of the quadcopter, as well as the tests carried out to check its operation.

CHAPTER 1

HISTORICAL CONTEXT ORIGIN OF UNMANNED AERIAL VEHICLES

This chapter describes the UAV development processes and their regulations.

1.1 First experiments with rotary wings

The creation of quadrotors was inspired by an attempt made by French professor Charles Richet in 1906. Richet had the idea of building a helicopter called Gyroplane No.1 with four support arms, 32 propellers, 8 on each arm, and a central motor to turn the propellers. Unfortunately, this project did not work out. After Professor Richet's failed attempt, his student Louis Bréguete and his brother Jacques Bréguete built the first quadrotor called the Bréguet-Richet Gyroplane No.1. The Bréguet-Richet Gyroplane No.1 took off in 1907, but it didn't have much stability and was only in the air for a short time. The Bréguete brothers used the same concept created by Professor Charles Richet, the Bréguet-Richet Gyroplane No.1 was equipped with a single 40/45 HP (30/34KW) engine, weighing the equivalent of 578kg. Professor Charles Richet's idea (1907) of flying an aircraft using rotating wings with propellers turning clockwise and anti-clockwise is the same principle used today in model aeroplanes with rotating wings.

Figure 1: Bréguet-Richet Gyroplane No.1

Author: https://alchetron.com/Breguet-Richet-Gyroplane-1969531-W

Since then, there have been more constructions aimed at rotary wings. In 1920, the QuadRotor Oemichen No.2, built by Étienne Oehmichen and weighing approximately 800kg, set a world flight record, staying in the air for 7 minutes and 40 seconds, covering a distance of 360 metres with little stability. After this achievement in 1921, the US Army hired Dr Georges Bothezate and Ivan Jerome to build machines that could lift vertically, but it didn't work out, their machines rose only 5 metres and weighed more than a tonne (SILVA 2014).

Figure 2: QuadRotor Oemichen No2 Author: https://alchetron.com/Breguet-Richet-Gyroplane-1969531-W

1.2First fixed-wing unmanned vehicles

Ever since the first wars in the 19th century, with the use of firearms, engineers have been thinking about creating ways of carrying bombs to blow up enemy territories without having to send soldiers to the areas of conflict. With the advance of technology in the arms industry, the main aim of creating missiles or unmanned aerial vehicles (UAVs) was to reduce the loss of life on battlefields. One example was the creation of the Fritz X1 missile (1938), hot air balloons (1849) and remotely guided UAVs (1935). Over time, these technologies began to be used, developed and improved for a variety of purposes, such as agricultural fields, leisure, the environment, surveillance and other areas that needed a good quality aerial monitoring vehicle that was low cost compared to aeroplanes, with low maintenance costs and high efficiency.

Two facts that marked the beginning of the use of UAVs were the use of cameras and the measurement of wind speeds at altitude using a flying artefact. According to Puscov (2002 apud Medeiros 2007), the term UAV has been used since 1883, when Douglas Archbald had the idea of installing an anemometer on a wire to measure wind speed at various altitudes until it reached a height of 1200 feet (equivalent to 365.76 metres).

On 20 June 1888, in France, Arthur Batat attached a camera to a pandora and recorded aerial images - a historic milestone (MEDEIROS, 2007). In 1709, the Brazilian priest Bartolomeu Lourenço de Gusmão demonstrated his project in Portugal, in the presence of King João V, which consisted of a manned hot air balloon that could be used as a means of aerial attack or just for leisure.

On 22 August 1849, Austrians used the Brazilian priest's idea to create a manned hot air balloon, with an innovation: they didn't need a crew to control the balloons, they would be used for bombing, and so they began experimenting with unmanned aerial vehicles. The balloons were guided by cables attached to their carcasses. With the hot air, they rose and were pushed only by the force of the wind. In this way, the wind dragged the balloon, loading it with time bombs, towards the target. Documents report that around 200 balloons were sent against the city of Venice in Italy as a response to the revolt of the inhabitants who had lost their territory to Austria (LONGHITANO 2011).

Figure 3: Air raid simulations using a balloon Author: https://static1.squarespace.com/static/52646bd2e4b04464b5400308/t/52bbfde6e4b0525234c4983d/

After the events of the creation of unmanned balloons and gyroplanes, which didn't work out very well because they were slow or unstable, the development of compatible designs and systems for a model aeroplane to fly without a crew started to become more possible from 1935 with the creation of the RP-1.

In 1935 Reginald Denny designed the RP-1 or RPV (Remote Piloted Vehicle), considered to be the first unmanned aerial vehicle. The RP-1 was controlled by a radio signal, had a fuselage to withstand artillery and flew only in a straight line.0 In 1938 Reginald Denny went to demonstrate the RP-1 for the UA armed forces, but the radio signal failed during the flight and the RP-1 ended up crashing.

From the RP-1, Reginald Denny began to develop more UAVs, such as the *RP-2, RP-3 and RP-4,* with various flight experiments. In 1939, the US Army ordered 53 UAVs from Reginald Denny, the most sophisticated of which was designated OQ-1. Modelled on the RP-1,

the RP-5 also appeared and was accepted by the US Army and Navy air services.

According to Hardgrave (2005), in 1938 a German company began creating gliding bombs guided by a rocket carrying 300kg of explosives under the name Fritz X to attack enemy ships. Hardgrave (2005 apud Medeiros 2007), soon after the OQ-1 came the OQ-2 and OQ-3 with many trials and tests to perfect them. In 1943, the first flight was made with the capacity to reach 165 km/h, using a fuselage made of steel tubes, using an O-15-3 engine with a single propeller, without a landing gear, as shown in figure 4 of the OQ-3.

Figure 4: OQ-3 image

Author: http://www.ctie.monash.edu.au/hargrave/images/denny_oq-3_elpaso_tx_1_750.jpg

According to Silveira (2005), the 1982 Lebanon War was the most important milestone in the development of UAVs. In that year, Israel managed to use a UAV to detect 17 of the 18 Syrian anti-aircraft batteries in the Bekaa Valley, with area reconnaissance carried out only through overflights.

After all these historical events, the US army began to invest more in this technology to eliminate attacks and territories with enemy dangers. In the mid-1990s it began using the UAV called the MQ-1 Predator. In 2001 it was used in the war against Afghanistan, carrying a missile that was successfully launched at enemy troops. The Predator is equipped with state-of-the-art equipment and its purpose is weapons reconnaissance, site reconnaissance for intelligence, surveillance and reconnaissance of enemy capabilities, and it can fly for up to 40 hours without landing (MILESKI, 2007).

Figure 5: Predator

Author: http://www.airplane-pictures.net/photo/168449/00-432-usa-air-force-general-atomics- aeronautical-systems-mq-1-predator/

In Brazil, the first UAV project was created in 1984 under the name Acauã IPD-8408. Its main objective was military and civilian applications, such as tactical reconnaissance with flights close to the ground, radar identification and natural resource sensing. Four metal carcasses and one wooden carcass were prototyped for this project. It was completed and flew for the first time in December 1985 for the purpose of area reconnaissance (BRANDÃO et al. 2007).

According to Oliveira (2005), the Aerospace Technology Centre (CTA), after developing the Acauã together with the development project to launch a missile called Piranha using a UAV, continued with various tests for the Brazilian Air Force to create a UAV that would launch a missile. After a decade and with technological development, the CTA continues to develop the UAV Project, but now with the initiative of the Ministry of Defence, with the aim of serving Brazil's three major armaments forces, i.e. the Brazilian Army, the Brazilian Navy and the Air Force. The respective services are aimed at: tactical and aerial reconnaissance, on-board reconnaissance (ships and platforms), aerial targeting such as aeroplanes and UAVs.

Figure 6: Acauã

Author: http://www.defesanet.com.br/aviacao/noticia/11709/

1.3 Requirements for owning your own drone

Unmanned aerial vehicles are known by the following acronyms: UAV (*Unmanned* Aerial Vehicle), RPAS (Remotely Piloted Aircraft Systems), UAV *(Unmanned* Aerial *Vehicle*). In addition to the acronyms created for unmanned aerial vehicles, it can also be called a *drone*. Its structure is generally in the shape of a quadcopter, but its name is colloquial and its translation would be "drone" because of the noise emitted by its engines when it takes off.

With the expansion of model aeroplanes in Brazil, the National Civil Aviation Agency (ANAC) and the Department of Airspace Control (DECEA) have created some simple rules for their use.

According to ANAC's Special Brazilian Civil Aviation Regulation No. 94/2014 (RBAC-E No. 94/2017), it complements the drone operation rules established by the Air Control Department (DECEA) and the Telecommunications Agency (ANATEL), such as:

> Respect the distance limits of third parties and observe the rules of DECEA and ANATEL. Model aircraft with a maximum take-off weight (including the weight of the equipment, its battery and any cargo) of up to 250 grams do not need to be registered with ANAC. Model aircraft operated in line of sight up to 400 feet above ground level must be registered and, in these cases, the remote pilot of the model aircraft must have a licence and qualification (ANAC 2017).

In addition to this rule, the drone driver must follow the class established by ANAC, which is divided into three classes:

Class 1 aircraft weighing more than 150kg must follow these instructions. "Aircraft must be certified by ANAC, registered in the Brazilian Aeronautical Registry (RAB) and pilots must have an Aeronautical Medical Certificate (CMA), licence and qualification. All flights must be registered" (ANAC, 2017).

Class 2 weighing between 25 and 150kg must follow these instructions:

> Aircraft don't need to be certified, but manufacturers must fulfil the technical requirements and have the project approved by the Agency. They must also be registered with the RAB and pilots must have a CMA, licence and qualification. All flights must also be registered (ANAC, 2017).

Class 3 weighing more than 250 grams up to 25kg must follow these instructions:

> If operated up to 400 feet above ground level (approximately 120 metres) and in line of sight, they will only be registered (presentation of information about the operator and the equipment). No CMA will be required, nor will it be necessary to register flights. Licences and permits will only be required

> for those wishing to operate above 400 feet. RPA operations of up to 25 kg can only take place at a minimum distance of 30 metres from a person. The distance may be less in the case of consenting persons (those who expressly agree to the operation) or persons involved in the operation. In urban areas and rural agglomerations, operations will be a maximum of 200 feet above ground level (approximately 60 metres), (ANAC, 2017).

UAVs have two classes of aeromodels, fixed wings (aeroplanes) with aerodynamic support from the air flow in their wings. Rotating wings (Drones) with thrust created by the propellers (ROSKAM, 2003). This technology has grown a lot, especially in the area of agriculture to provide information to the agronomist or farmer himself about his crops, while for cities focused on infrastructure in large harbours and huge buildings, drones have become a good business for filming and taking aerial photos at a low cost to the consumer and with excellent quality.

For these benefits, rotary-wing UAVs are widely used: Tricopters, Quadcopters, Hexacopters and Octacopters, which are supported by propellers attached to the frame's engines. This denomination is given by the number of engines that will be used to have propulsion force for the use of the UAV, the most used are with 4 engines carrying a lower amount of weight than the other UAVs, then with 6 or 8 engines it will present a superior force with this it can handle much more weight, but its energy consumption increases (DEMOLINARI 2016).

In this project, the choice was to build a quadcopter-type drone with four propellers, good stability, accessible and cheap parts and easy to manoeuvre during flight.

CHAPTER 2

QUADCOPTER STRUCTURE

This chapter describes the main parts and components used to assemble the quadcopter.

This description is necessary to provide an understanding of the individual components and their function in the overall operation of the quadcopter.

2.1 Quadcopter construction diagram

The construction of the Quadcopter follows the diagram in Figure 7.

This diagram shows where each part and sensor used is located. This diagram can be used as a reference for assembling quadcopter drones to offer services in leisure, commercial, security, rural and environmental areas, as it brings integrated systems and sensors that offer stability and agility in carrying out their tasks (SANTOS, 2016).

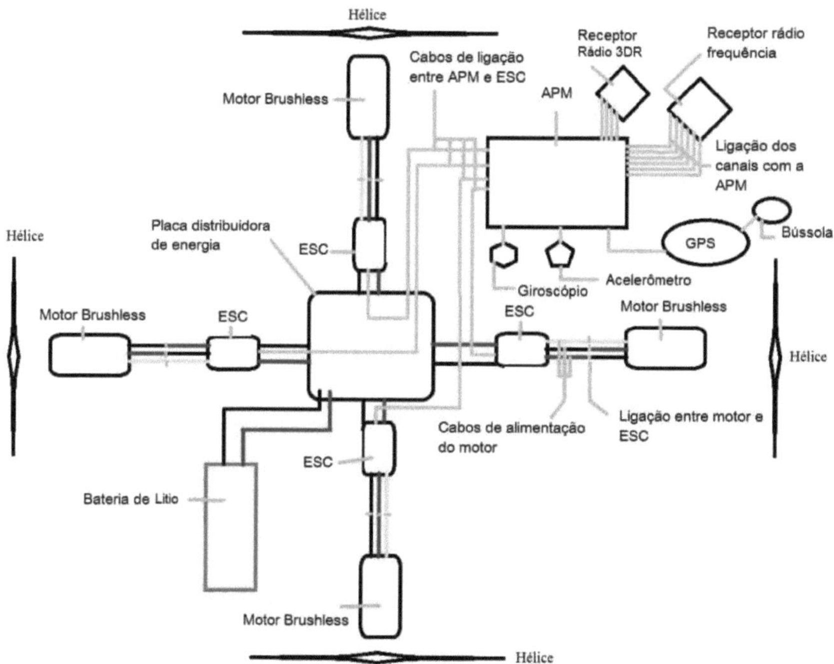

Figure 7: Quadcopter diagram Author: Prepared by the author

2.2 Flight dynamics

Drones have slightly different flight dynamics to aeroplanes and remote-controlled aircraft, but they are similar to helicopters in that they fly vertically and horizontally with easy stabilisation on their rotating wings. The movement and stabilisation process is carried out using the thrust provided by the propellers and motors, which work to orientate the drone in its three dimensions of Pitch, Roll and Yaw, which is passed on by the radio control signal.

Pitch is the forward and backward movement of the Drone, made by the lever (or stick aileron) of the radio control located on the right of the radio.

control, if you move the quadcopter upwards it goes forwards and if you move it downwards it goes backwards.

Roll makes the quadcopter fly sideways, to the left or to the right, if you move the *aileron stick* to the left, the quadcopter will fly to the left, if you move the aileron stick to the right, the quadcopter will fly to the right.

Yaw causes the front of the quadcopter to turn to the right or left, and can be controlled via the throttle lever by pushing left or right (ROMERO, L; POZO, D; ROSALES, J; 2014).

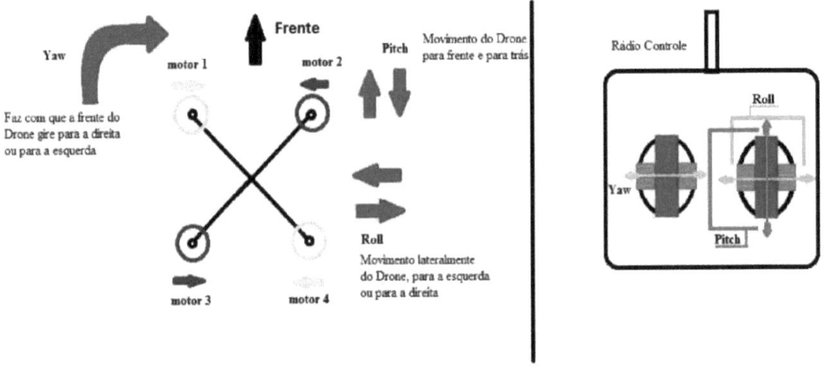

Figure 8: Simulation using Yaw, Pitch and Roll Author: Prepared by the author

As seen in Figure 8, the quadcopter is able to perform its tasks in three dimensions and for this to be done the motors need to increase and decrease thrust, the execution of the motors will be as follows (SANTOS, 2016).

Pitch flight forwards: Motors 1 and 2 slow down and motors 3 and 4 speed up.

Pitch flight backwards or return: motors 1 and 2 increase speed and motors 3 and 4 decrease.

Roll flight to the right: motors 2 and 3 slow down and motors 1 and 4 speed up.

Roll flight to the left: motors 2 and 3 increase speed and motors 1 and 4 decrease.

Yaw by turning the front of the quadcopter to the left: motors 2 and 4 increase speed and motors 1 and 3 decrease speed.

Yaw by turning the front of the quadcopter to the right: motors 2 and 4 slow down and motors 1 and 3 speed up.

The quadcopter has these three dimensions plus a speed control called a throttle that makes the quadcopter increase in altitude, decrease and stop at a certain distance with the four motors rotating at the same time at certain speeds (SANTOS, 2016).

2.3 Brushless motor

The motors used in remote control cars and UAVs fall into two categories: brushed and brushless. The brushless motor, which is the most widely used, can be characterised as modern: it uses permanent magnets in its bases, and its mechanical part can be compared to a motor with brushes. The Brushless motor has no contact between the stator and rotor for electricity to pass through, its structure is made up of between four and eight magnets on the outside, enabling it to rotate on its own axis and create the speed required for its rotation, unlike brushed motors, which have their coils fixed in their structure, the three power cables that come out of the Brushless motor with different colours that are connected directly to the ESCs (FOUR, R.; RAMOUNTAR, E.; ROMEO, J.; COPELAND, B, 2007).

When the motor receives the electrical pulses sent by the ESC controller, their value alternates between pulse voltages 1 and 0: 1 positive and 0 negative, these pulses will generate in the coil a north electromagnetic field like a magnet, the electromagnetic field of the coil will expel with the magnetic field of the magnet that is located in front of the coil as in figure 9, the magnetic fields of the magnet are North and South, when two north fields conflict and tend to expel each other, causing the coil to rotate towards the next magnet, magnetisation will occur in all the coils to turn the motor, this type of motor uses around 12 coils to be electromagnetised (FOUR, R.; RAMOUNTAR, E.; ROMEO, J.; COPELAND, B, 2007).

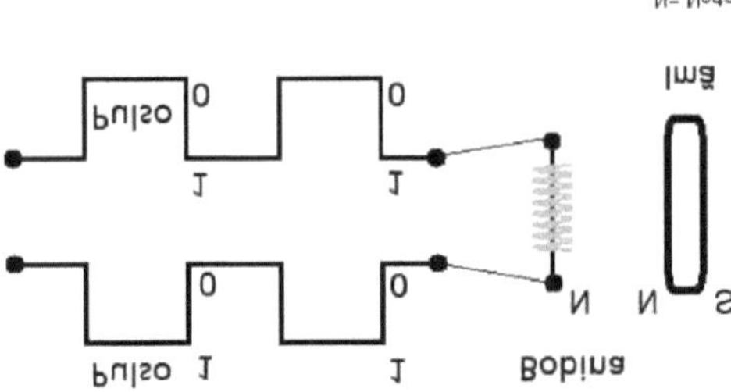

Figure 9: Brushless motor cycle Author: Prepared by the author

The internal connection of brushless motors works in two ways, as there are three different coloured wires identified as black, yellow and red, which are connected to the caps as shown in figure 10 (FOUR, R.; RAMOUNTAR, E.; ROMEO, J.; COPELAND, B, 2007).

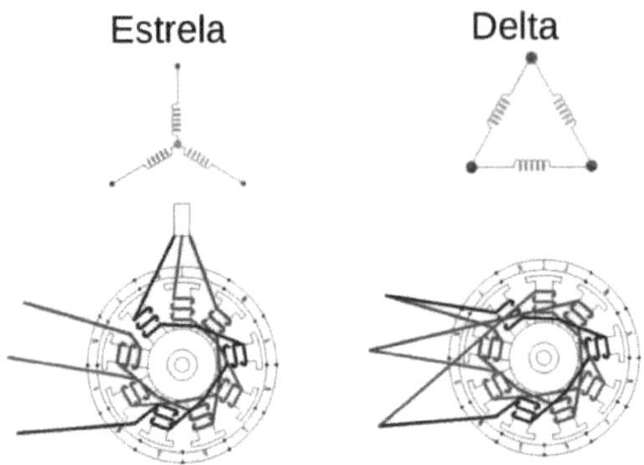

Figure 10: Brushless motor
Author: http://www.c2o.pro.br/automacao/figuras/bldc_config_estrela_delta.jpg

These connections are made in the factories themselves and are supplied as standard for use in robot intuition and areas that require a brushless motor. When the motor is connected in star, it consumes less energy than in delta because it has less turning power. The delta connection has the same current for all three coils that are sent by the ESC pulse and has a higher speed than the star connection (FOUR, R.; RAMOUNTAR, E.; ROMEO, J.; COPELAND, B, 2007).

Brushless motors operate with DC (Direct Current) which has a high power and good torque performance, this helps their flight and as it is a brushless motor the useful life is long, the motor has an ideal size for the drone's structure (FOUR, R.; RAMOUNTAR, E.; ROMEO, J.; COPELAND, B, 2007).

The drone parts market has many options. You can choose a motor with a high rotation capacity, medium rotation and low rotation, but its price is also quite variable, depending on its rotation capacity. For this quadcopter structure it was decided to choose the Turnigy D2836/8 1100KV (rotation/voltage unit) Brushless Outrunner motor with medium rotation capacity figure 11.

Figure 11: Engine used to realise the project Author: Prepared by the author

To start up the brush motors, the electrical circuit was used in the four motors with the following connections: two with the wires (black = black, red = red, yellow = yellow) and two with wires of different colours (black = red, red = black, yellow = yellow) in the speed controllers (ESC), enabling the motor to rotate correctly with the propellers as shown in figure 12 to create thrust in order to cancel out the weight of the central axis (SANTOS, 2016).

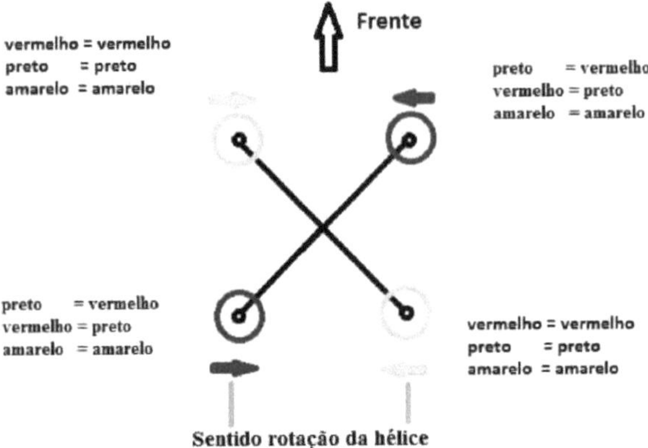

Figure 12: Simulation of the wires to be connected Author: Prepared by the author

2.4 ESC (Electronic Speed Controller)

The ESC controller (Electronic Speed Controller) is one of the indispensable tools for this project, as it gives the motor its correct torque.

The ESC controller operates by means of the signal sent by the APM flight controller and connections to the power distribution board to establish all the pulse signals on the brushless motor. The connections made on the ESC board to their recipients are made as follows: at the controller's input we have the red wire (+ positive) which receives power from the battery at a DC voltage of 5.5 to 16.8V together with the black wire (- negative) which are connected directly to the power distribution board.

On the other side of the ESC controller there are three wires that feed the motor: black, yellow and red, all connected directly to the motor. The thinnest wire with different colours will be used to receive the PWM (Pulse-width modulation) signal from the APM board and direct it to the motor for it to turn (ZACCARELLE, 2012).

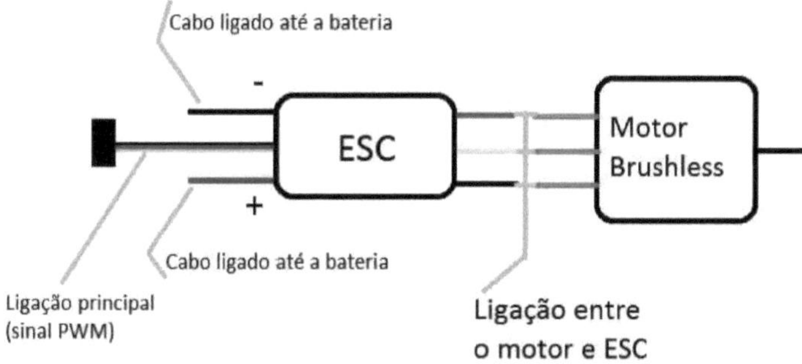

Figure 13: ESC's main link Author: Prepared by the author

The ESC controller chosen for the project is a 20 A AFROESC brand (Figure 14) with all its electronic circuits protected with a layer of plastic to prevent them from getting wet or dirty. This project used four models connected to each brushless motor.

Figure 14: ESC
Author: Prepared by the author

2.5 Hobby King kk 2.1 controller board

The HobbyKing KK2.1 multi-rotor flight controller is designed to control multi-rotor UAVs such as Tricopters, Quadcopters, Hexcopters, etc. The purpose of this board is to stabilise the aircraft during flight by taking signals from the gyros to its processor, which processes the signals according to the users. This controller works together with the speed controllers (ESCs) to manage the rotational speeds of the motors, which in turn stabilise the aircraft. There is only one way to control this board, which is via radio control signals and its stabilisation mode is performed via the aileron and elevator.

Once processed, this information is sent to the ESCs which, in turn, adjust the rotational speed of each motor to control flight orientation.

2.6 Controller board and its components

In this work we chose the *ArduPilot Mega* (APM) board, an electronic circuit board with Open Source programming code dedicated to achieving inertial navigation through its sensors. The board's platform was based on the *Arduino* development environment to make it easier to use and modify the programme, which is supported for use in boats, fixed-wing and rotary-wing aircraft, helicopters and carts, all of which are guided by radio control or telemetry.

The APM flight controller has been developed since 2009 with its first version called *ArduIMU.* With the development of the first controller board, it became possible to write other codes and improve its software and hardware structures. The best known and most used versions are APM 1 and APM 2, developed for UAVs. Version 1 was developed in 2010 with two electronic circuit boards: the red APM board and the blue IMU *(Inertial Measurement Unit).* The second version, APM 2, was developed in 2011 with more sensor support and everything coupled to its hardware, which is protected against damage to its electronic components.

The APM 1 board, as mentioned above, is connected to two boards: the first (red) board is allocated to the processor and co-processor for connecting and communicating with the radio and other components; it also has USB, GPS and *DataFlash* inputs. The second board (blue), which is above the red one, is used to connect sensors such as the gyroscope, accelerometer, barometer, pressure and voltage sensors. This version of the board has a different power input to the APM 2.

APM 2 is a board that contains microprocessors with 8 input ports and 8 *PWM* output ports and has other ports (to which new sensors can be added), it has several inbuilt sensors, such as a barometer, magnetometer, gyroscopes in x, y and z, 3-axis accelerometers. The board's function is to process the signals received from the receiver, together with the data from the 13 other sensors, and send the signals that will move the motors as desired to the output (BARATO 2014).

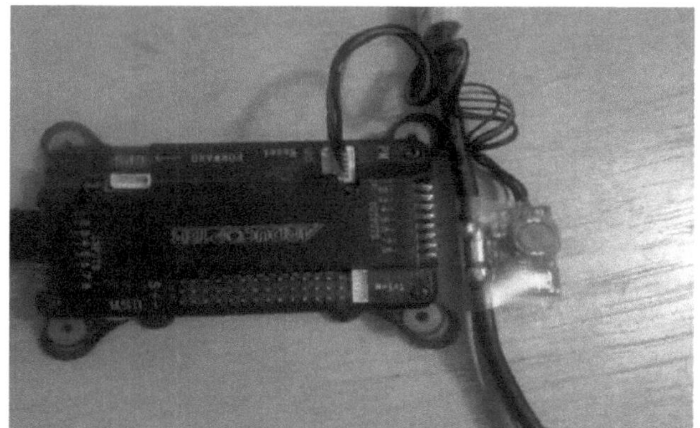

Figure 15: Controller board Author: Prepared by the author

The APM board has 8 pin inputs on the left that receive the signal from the radio control receiver and 8 outputs on the right to transmit the signal to the ESCs. At the top bottom of the board there are two GPS signal inputs with an arrow pointing to them, and next to them the Power Mode signal input and output, while at the top there is the input for the telemetry device and 15 inputs or outputs for other accessories such as cameras and other sensors.

2.7 Mission Planner Software

Missio Planer is free open-source software developed by Michael Oborne to control the *ArduPilot Mega* APM board. The software is compatible with Windows and Apple computers. With the software we can configure all its sensors, tune the model aircraft, plan routes with the help of GPS, monitor the status of the quadcopter when it's in operation, among other functions.

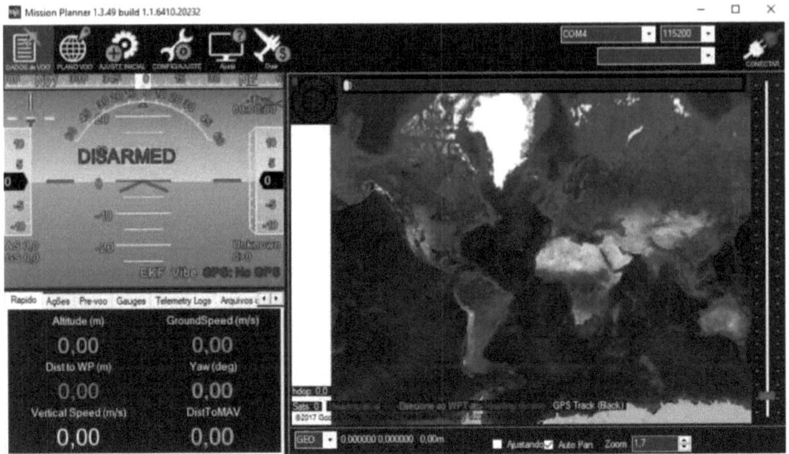

Figure 16: Mission Planner software Author: Prepared by the author

When the software starts, it opens the screen in the image above with many names translated into Portuguese to make it easier for the user to recognise each option. When the user connects the APM card to the software, it will automatically check that your *firmware* is up to date with the latest version of *Mission Planner*. If it isn't, it will give you the option to update and always get the best performance.

2.8 Propellers

Brushless motors need to work directly with the propellers to convert the mechanical work of the motor into propulsive movement for flight. The quadcopter works with four types of propellers, two with thrust rotation to the left and two to the right, as described in figure 12.

The image below shows the shape of the left- and right-facing propellers used in this project.

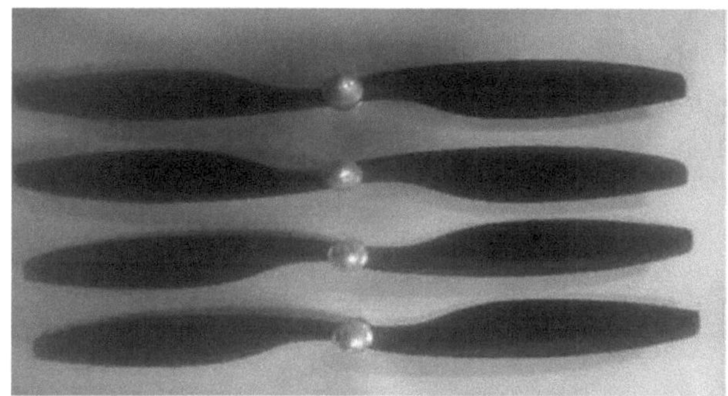

Figure 17: Propellers
Author: Prepared by the author

2.9 Battery

The battery is used to power all of the UAV's electronic circuits. It is the heaviest accessory, so its weight can unbalance the drone if it is placed in the wrong place; ideally, it should be positioned in the centre of the UAV, thus passing the same weight load to all the motors. The capacity of this battery is measured in milli ampere hours (mah) in our case it is a 2200mah Lipo Turnigy model that has the capacity to fly for around 10 to 20 minutes. The energy to be supplied by the battery to the electronic components is made through a connection to the Power Module and then the power distribution board that will support the drone (SANTOS, 2016).

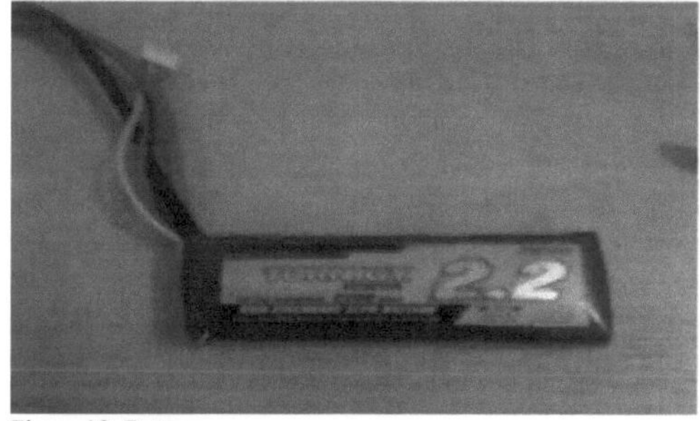

Figure 18: Battery
Author: Prepared by the author

2.10 Sensors

As described in Figure 08 at the beginning of Chapter 2, the APM flight controller board is made up of various sensors, some internal and others external, which help to ensure a more precise flight with good stability. Below we will describe these sensors, which are located on this quadcopter and inside the APM board.

2.11 GPS

The Global Positioning System (GPS) will be used to create flight routes without the need for a radio control just by using the telemetry command made by the Mission Planner Software. The GPS system provides the three-dimensional location with the time and speed to reach a given location.
The GPS navigation system is the most widely used system for easily accessible locations and positioning. GPS was developed by the US Department of Defence to be used by the American armies in times of war for land positioning and marking territory. In the 1970s it was improved with new technologies and began to be used by the general public.

According to Alberto (2011), the GPS system is basically made up of three segments: the spatial segment, the control segment and the user segment.

The space segment is localised as a constellation of satellites (twenty-four satellites) distributed in six orbital planes to generate the on-board navigator transmission.

The control segment is conferred by ground stations through satellite monitoring, control and maintenance.

The user segment is basically the GPS receivers, processors and antennas that make it possible to receive the radio transmissions used by civilian and military user communities.

Vettorazzi et al (1994) point out that the positioning and time of the location is done via the satellites to the GPS receiver, both of which use an internal clock to mark the time in nanoseconds when the signal is emitted from the GPS and received by the satellite. The calculation is made using this information, taking the timed time it takes to get from the receiver to the satellite. This information reaches the satellite more precisely because it works in triangulation and takes an average of the distance and time travelled, determining the exact position and your location.

The GPS is installed on the APM board and is recognised when the Mission Planer software map is run, giving its location and positioning. The function of the GPS on the drone will be to follow a flight route established by the software and to be executed accurately using the telemetry signal. In the event of a loss of signal, it will help the drone to return to the starting point and land or execute the mission given by the software and then return to the starting point and pose so that there are no accidents with the drone.

2.12 Gyroscope

The angular movement made by the quadcopter passes a direction position that is obtained by using the gyroscope, following the principle of Inertia of Newton's first law of physics "By which a body in motion remains in motion until a force external to it alters its motion". As the quadcopter moves, the gyroscope will pass the direction reference of its location with the help of the GPS device and pass its location or follow a route given by the Mission Planner, the movement is worked with the x, z and y axes with angular position control on the yaw, roll and picitch axes (VIEIRA, 2011).

2.13 Accelerometer

Accelerometer is a sensor used on the APM controller board to calculate and measure the acceleration of the quadcopter during its flight, measuring and controlling the pressure of static forces or gravitational forces that are caused by the drone's movement or vibration of the accelerometer (SANTOS, J. TEXEIRA, L. 2013).

2.14 Barometer

The barometer is a sensor used on the APM controller board that will calculate the drone as it moves away from the ground to its desired height with a calculation related to atmospheric pressure. This sensor works together with the GPS and has to work with perfect efficiency if it doesn't lose height and pass on the wrong information (SANTOS, J. TEXEIRA, L. 2013).

2.15 GPS Magnetometer and Compass

The Magnetometer is an instrument used on the APM controller board to measure the intensity, direction and direction of magnetic fields near it. The way it works can be explained as follows: the coils of the electromagnet receive an electric current, which then generates a magnetic field in the area - this area is magnetised and will produce an induced field that will be read by the sensor and converted from the magnetic field into an electrical signal (ANDRADE, 2012).

As such, this sensor works directly with the APM card to ensure that it is always positioned and that it knows the direction of the card in relation to the drone and its position. The compass will work together with the GPS where it is installed, but rather by passing on the coordinates to the Mission Planner software on the GPS positioning by measuring the Earth's magnetic field and passing on the magnetic north-south alignment and locating one point to the other (ANDRADE, 2012).

2.16 Inertial measurement unit - IMU

The inertial measurement unit is nothing more than a combination of components to help stabilise the UAV during its flight. Sensors such as the accelerometer, gyroscope and magnetometer will make it possible to determine the speed of the motors and position, guiding the quadcopter to maintain itself even when suffering changes in wind and gravitational forces. The inertial measurement unit (IMU) aims to determine an object's velocity from its acceleration and its angular position from its angular velocity (BARATO, 2014).

2.17 Radio control and 3DR radio

The project was developed with two types of remote control, one made via 3DR Radio and Radio Control.

3DR Radio version 2 is a telemetry connection to the APM card in a ground station. Its platform is open source, easy to use and has a wide range.

The 3DR Radio set includes the transmitter, which is connected to the computer using the Mission Planner software with the appropriate configuration, and the receiver, which is connected directly to the *APM* board with two *LEDs* signalling their status, one red and the other green with the following meanings: green LED flashing = radio searching, green LED on = radio concession, red LED flashing = data transmission, red LED on = firmware data update mode.

Turnigy radio control is the second remote control to be widely used in aeromodelling. It has five channels to be used as required, its channels result in pulses of between one and two milliseconds, its frequency is 2.4ghz with a voltage of 6V (four AA-size batteries).

To communicate with the radio control, you need a five-channel receiver that comes with the radio control and has a voltage of 4.8 - 6V. It is connected to the APM board and when it receives the signal from the radio control, a green LED will light up, otherwise it will flash and the configuration or signal is not being stabilised. In order to switch on the UAV, the joystick must be pressed diagonally towards the middle of the joystick. When it receives the signal, the ESC will beep and light up.

Transmitter Description and Setting

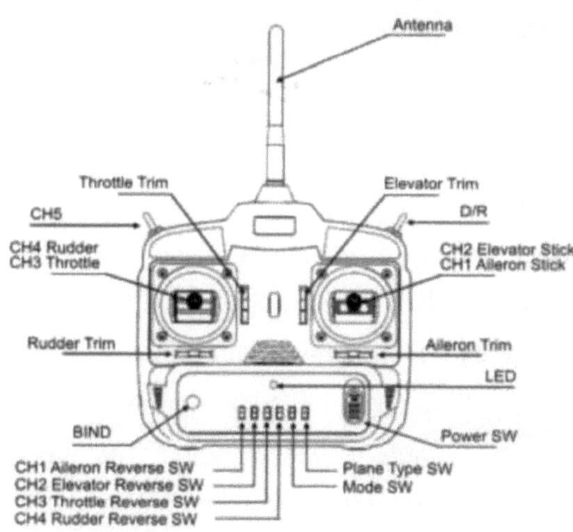

Figure 19: Radio control

Author: Radio Control Manual Guide

2.18 Chassis

In this project, two boards that are part of the drone were used to distribute the energy. These boards are of different sizes, as shown in image 20. The larger board is located underneath the drone and serves to provide sustainability and distribute the energy, while the smaller board is located at the top of the frame's arms to provide support and serve as a socket for the APM board. The larger power distribution board will supply the battery and distribute it to all the electronic components of this quadcopter through tin solder connections in the copper-coated parts. The wire that will be used has two colours, black and red with the following description: black for all negative connections (-) and red for all positive connections (+) (SANTOS, 2016).

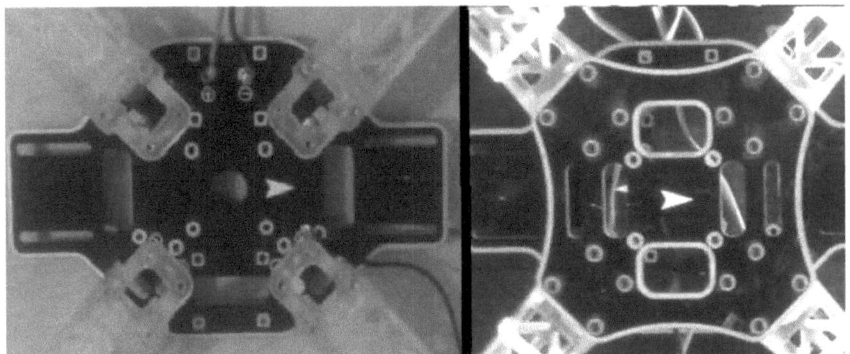

Figure 20: Energy boards Author: Prepared by the author

2.19 Frame

The quadcopter is built using frames known as drone arms, which are used to hold the motors and provide aerodynamics. Each arm has 450 mm between the power plate and the motor, providing a safe distance so that each propeller and other components are not hit during flight. Its weight with the plate is around 295 grams. The mode chosen for our Frame is the Q450 with its arms constructed from high-quality fibreglass and ultra-durable polyamide nylon. This model is reinforced, making it stronger and preventing it from breaking.

2.20 LED signalling

It has twelve green and twelve red LEDs that were used to help during its flight, depending on the distance the driver can see which is the front and which is the back of the drone (SANTOS, 2016).

2.21 Power Module

This accessory provides a stable power source and reduces the chance of a quadcopter blackout by monitoring battery voltage and current in real time, indicating to the software how much battery is left in the drone.

CHAPTER 3

PROJECT STAGES AND RESULTS

This chapter presents the order in which the drone was assembled. In other words, how it was built, the stages and the results. We will also present how the components were connected, how the software was installed and how the drone was calibrated together with the software. And also the first tests of the drone.

3.1 Drone design

The drone project began with the sizing of the parts to be used. The following list of parts and components was generated from this sizing.

- Frame and power distribution board set: a *Diatone* Q450 V3 Glass Fiber frame board was chosen because it is very popular in the hobbyist world and allows for simple and easy assembly.
- *Brushelss* engines*:* the *D2836/8* model was selected because it provides a thrust of 1130 kilo gram force.
- Propellers: selected according to the engine used.
- Battery. The *Lipo Turnigy 2.2* model was chosen taking into account its capacity (15 to 20 minutes) its weight and its value (188 grams, 100 reais) as it is only a prototype, there is no need for a very elaborate battery at this stage of the project.
- Charger: comes with the battery, sized specifically for the *Lipo Turnigy* model.
- LED: choice of two colours green and red for drone location and direction.
- ESC *(Electronic Speed Controller)* selected according to the current of the motor used, we chose the *AFROESC* with a capacity of 20 Amps.
- GPS *(Global Positioning System),* selecting the Ublox M8N model

because it has better accuracy and incorporates an internal compass.

- APM *(ArduPilot Mega)* A model 2.6 board was selected with the necessary characteristics for configuration and flight execution.
- *Power Module:* the XT60 model was selected because of its current-carrying capacity.
- Radio control: a simple radio is used, just to allow basic control of the drone. The *Turnigy* 5X model, a 5- or 6-channel radio, already meets the project's needs.
- *Mission Planner software,* provided by the APM flight controller website, provides all the functions needed to fly the drone.

With the list of parts and components needed for the quadcopter, we set about researching the following choices:

The first item on the list to be researched was the frame that provides the main support for the drone, with four support arms and a resistant material. The best-known brand for building a quadcopter is *Diatone*'s Q450 V3 model, which is drop-resistant in the event of a fall during the testing phase.

To illuminate the arms of the frame during night flights, and to locate the drone and its direction, we placed 4 green and red LAD plates indicating the green light in front and the red light at the back of the drone.

We then searched through scientific articles and forums with specific testimonials from experts on the motors used in quadcopter drones to find out which would provide the best performance and durability for short flights. We then decided to choose the Turnigy D2836/8 motor for our quadcopter, which is the most recommended for its performance, its screw fitting into the drone arm is standard, this model of motor can pull air up to around 2kg, reaching approximately 11000 RPM (rotation per minute) in each motor at its maximum speed.

When choosing the engine and its power, we had to decide on the size of the propeller, which is given in centimetres, with the most commonly used sizes being (7x3, 7x4, 7x5, 8x4, 8x5, 9x5, 9x6, 10x5, 10x6, 10x7, 11x4.7, 11x6, 11x7) and other approximate numbers. So we had to choose an ideal size so as not to overload the engine, which in our case was 10x5, which is the most recommended for this type of engine. If the propeller size you choose is ideal for your engine, it will make the engine more efficient: above the size you'll get more power, but more consumption; below the size you'll get more speed, but very little thrust.

The choice of battery was made together with its charger, which during the research proved to be the most sought-after among drone developers, as it can withstand flights of approximately 15 to 20 minutes, powering all the components without suffering any loss of energy. Its capacity is 2200 mAh and it weighs around 135 g.

The ESC was chosen according to the motor's power to perform the necessary thrust rotation with a current consumption of 12 Amps and a weight of 10 grams.

Our first idea was to create a code on the arduino board using GPS, but our research showed us that there was a board that had these characteristics. The APM board is open source so you can change and add new codes with a language similar to arduino and it has access to GPS.

During the search for the best board to use in this project, we had in our favour some components that were already installed on it, such as a gyroscope, accelerometer, barometer and pressure gauge to help with flight performance.

After defining the board, we chose the GPS and Power Module, both of which communicate easily with the software and the APM board and are the most recommended for use with this board. Finally, there were two ways of communicating with the flight controller: radio control and radio telemetry communication, passing the command through the computer via the free Mission planer software.

3.2 Assembling the drone

The assembly process was carried out by separating all the parts and screws from the quadcopter, starting with the frame, LED and motor. To connect one arm of the frame to the LED and motor, four screws were used on the underside of the arm for this connection.

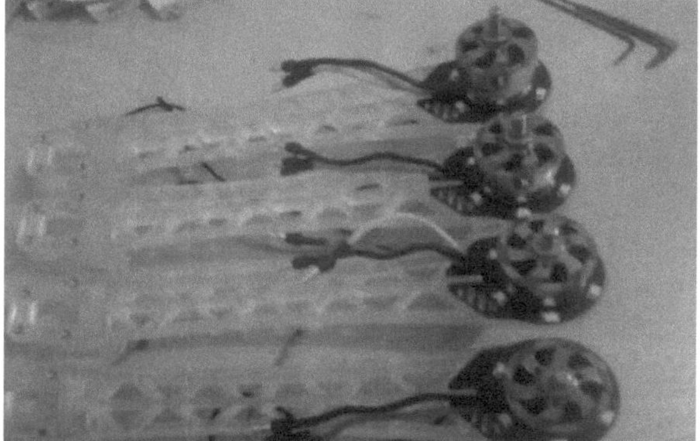

Figure 21: Assembling the Quadcopter Author: Prepared by the author

The next step is to connect the LEDs to a power source to see their colours and place them in the correct position, with green = green for the front and red = red for the back of the drone.

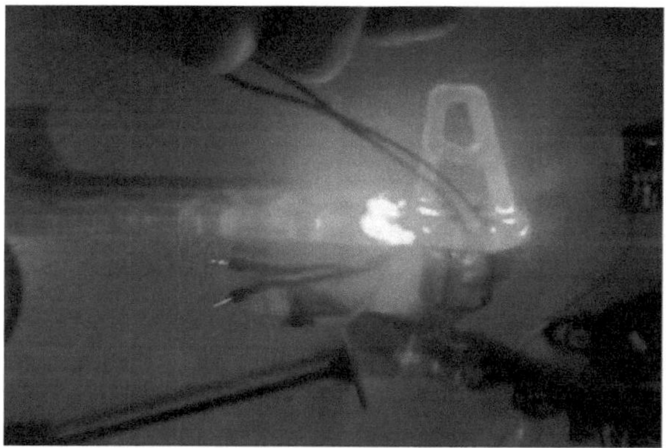

Figure 22: Assembling the Quadcopter Author: Prepared by the author

Once the frame is correctly positioned, we connect the arms of the frame to the power board and solder the two positive and negative cables that will carry the power from the battery.

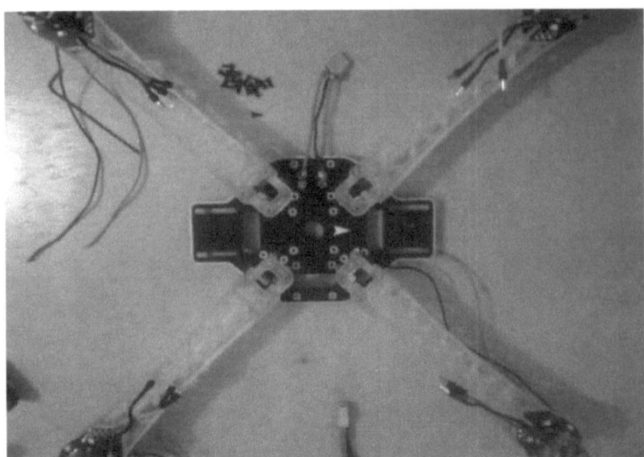
Figure 23: Assembling the Quadcopter Author: Prepared by the author

After connecting all the arms to the board with screws, we connected the ESCs to each motor and soldered them to the power distribution board. The next step was to attach the plastic clamp to secure the ESCs to the arms and connect the smaller power board to the frame.

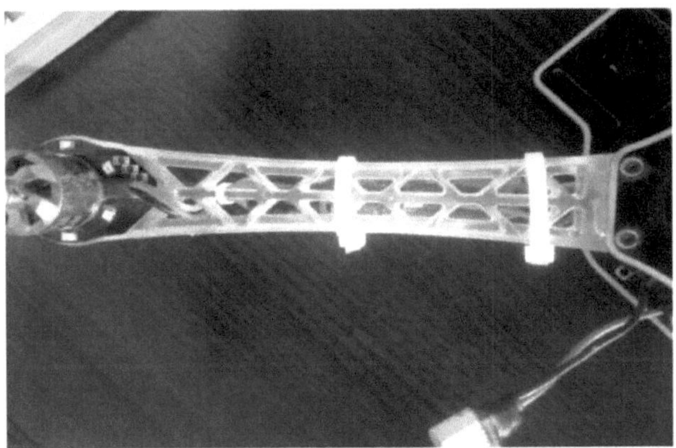
Figure 24: Assembling the Quadcopter Author: Prepared by the author

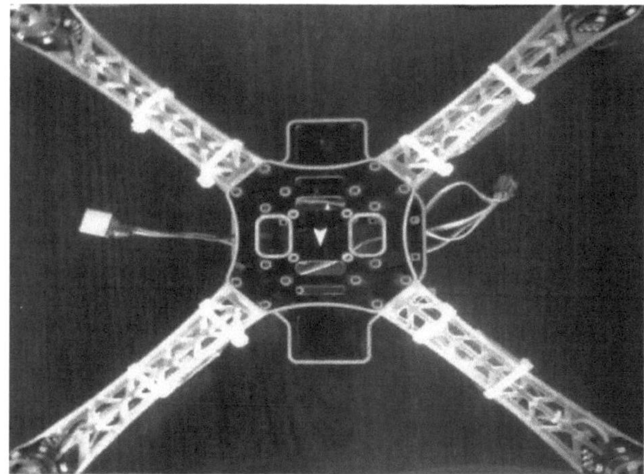

Figure 25: Assembling the Quadcopter Author: Prepared by the author

With all the connections soldered and secured, the next step was to add the GPS, APM board and radio receivers. Finally, we connected the wires between the system and connected the battery to the distribution board.

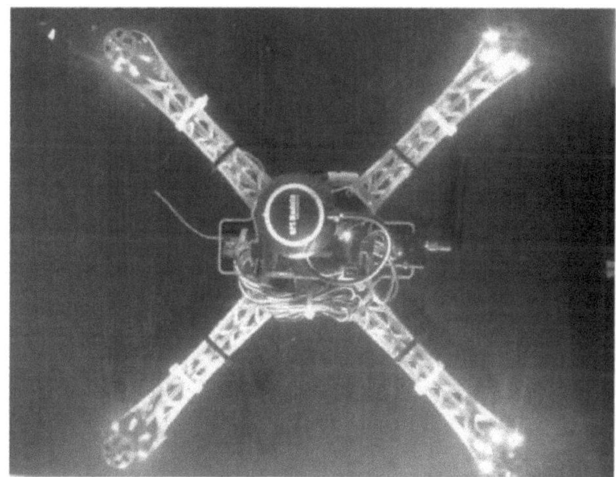

Figure 26: Assembling the Quadcopter

Author: Prepared by the author

3.3 Installation of the apm software (Mission Planner)

When you download the Mission Planer software from http://ardupilot.org, install it using the following steps: First step is to open the file and select the "Next" option to safely release the programme installation.

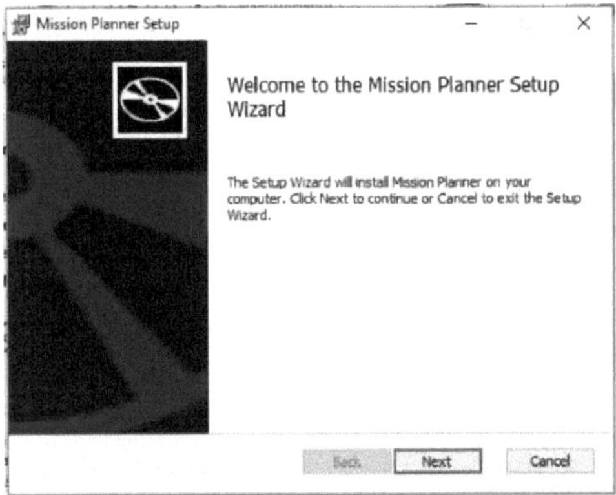

Figure 27: Installation Author: Prepared by the author

After running, a new window will open asking you if you agree to the *Software* terms, select the option and press "Next".

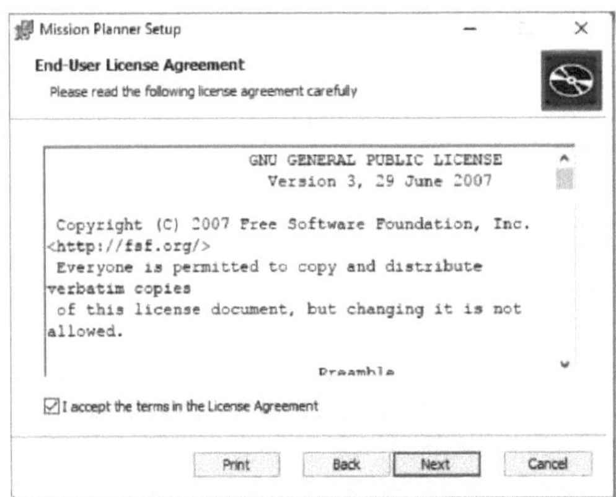

Figure 28: Installation

Author: Prepared by the author

Select the directory where your file will be saved and continue with "Next".

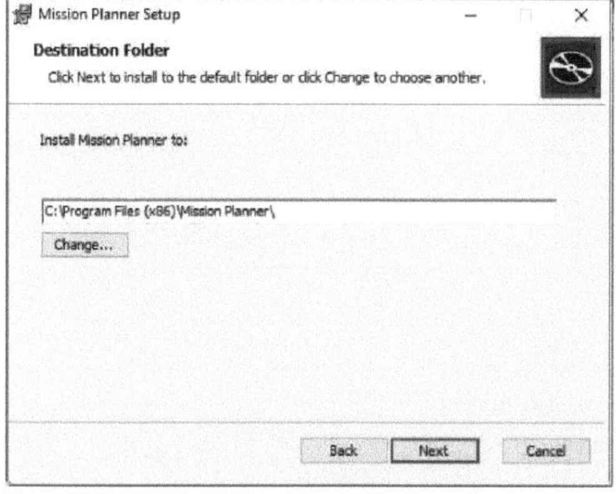

Figure 29: Installation Author: Prepared by the author

To continue the installation, select the *"IsstalF"* option to run the installation.

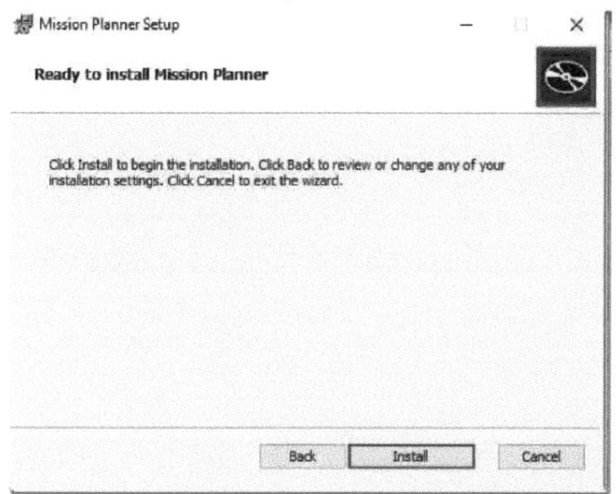

Figura 30: Installation
Author: Prepared by the author

After installing the Mission Planer Software, the programme will ask you to run the Drive, click on next until it runs.

Figure 31: Installation

Author: Prepared by the author

Once the program has finished running, it is installed.

Figure 32: Installation Author: Prepared by the author

3.4 Getting started and configuring Mission Planer

Once you've built the quadcopter and installed the Mission Planer software, connect the controller to the computer to open the software. The first requirement is that you have internet at your premises so that you can update the board directly from the developer, but if you don't want to there's no problem, but you should always keep the software up to date. Once the software has been updated, you can open the map that comes installed in the programme, then check that the software has found the input port in the top corner of the screen (figure 33) if it hasn't, just click on AUTO, you can leave it at 1200 and click on connect.

Figure 33: Mission Planner software Author: Prepared by the author

On the main screen when you connect, Google Maps will open with a drawing of the drone and informing you of the GPS direction with the position of the red line in front of the drone. On this first page, the software will provide information on the levelling of the APM board related to the drone as it stabilises on the ground, as shown in figure 34.

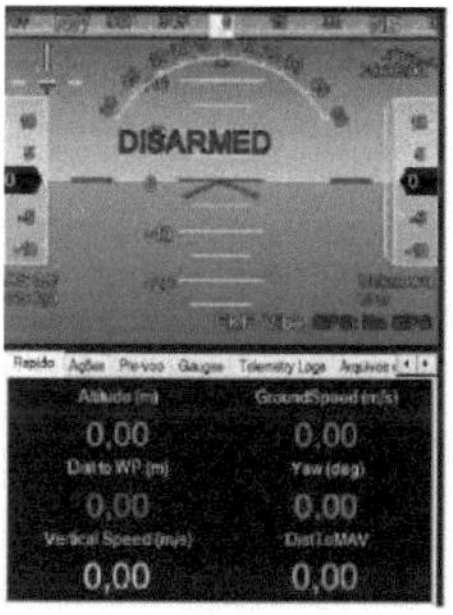

Figure 34: Mission Planner software Author: Prepared by the author

Figure 35: Mission Planner software Author: Prepared by the author

To make the first settings in the software, simply click on Initial Settings Wizard and select the model type of your UAV.

Figure 36: Vehicle model Author: Prepared by the author

The quadcopter model is a Multirotor, select it by clicking on it, then click on next, a new window will open showing the multirotor formats that the APM card can configure figure 37.

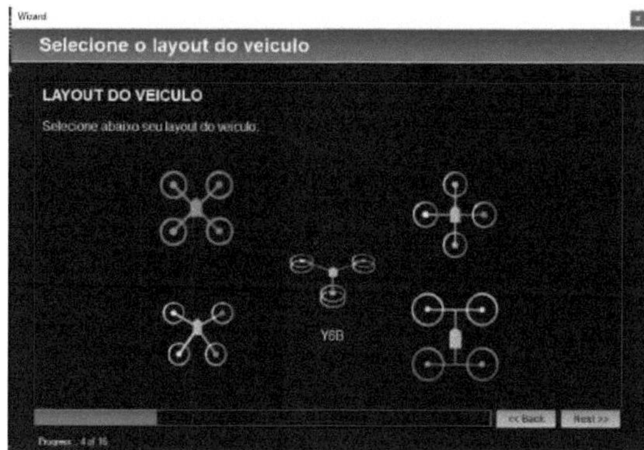

Figure 37: Drone layout Author: Prepared by the author

This project has been chosen to build an X-shaped quadcopter. Select the *layout* format *and* click "Next" to continue.

After choosing the model and its layout, you'll be asked to calibrate your accelerometer. This stage is very important for the quadcopter as it will calculate the speed and stabilise the drone, so it can't change position before you press continue. To configure, click on the green "start" button and make the quadcopter as shown in the picture stable and click "Start".

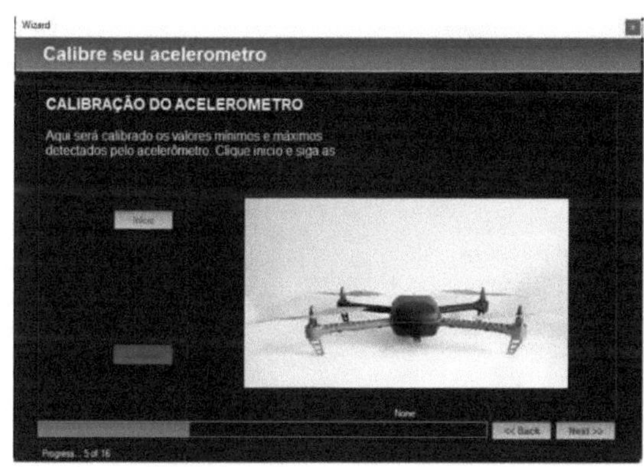

Figure 38: Accelerometer calibration Author: Prepared by the author

The next step is to turn the quadcopter to the left, the right-hand propellers up and the left-hand propellers down as shown in the image and click on "continue".

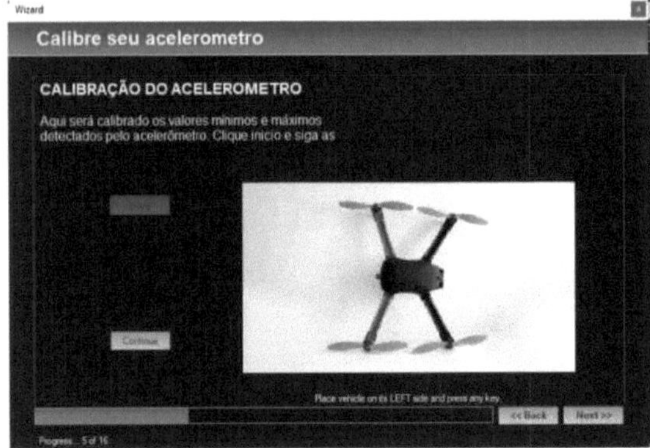

Figure 39: Accelerometer calibration Author: Prepared by the author

The next step is to turn the quadcopter right side up, the right propellers down and the left propellers up as shown in the image and click on "continue".

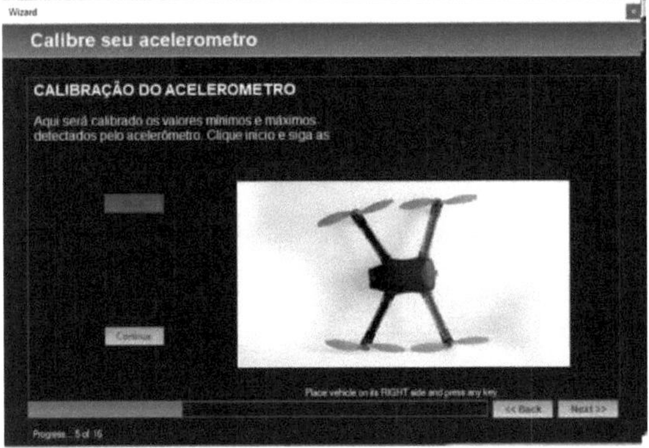

Figure 40: Accelerometer calibration Author: Prepared by the author

The next step is to put the front of the quadcopter down and the back up as shown in the image and click "continue".

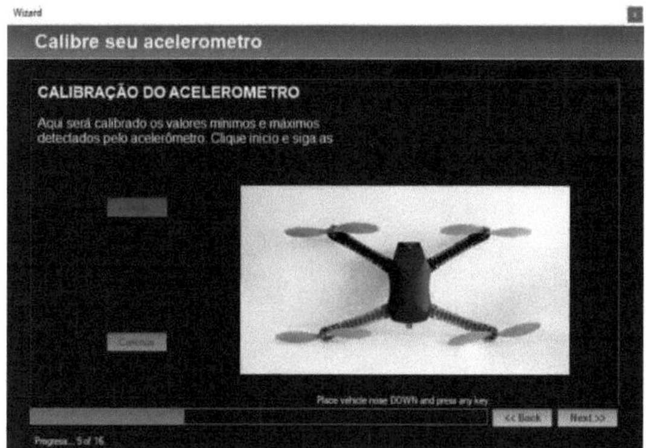

Figure 41: Accelerometer calibration Author: Prepared by the author

The next step is to put the front of the quadcopter up and the back down as shown in the image and click "continue".

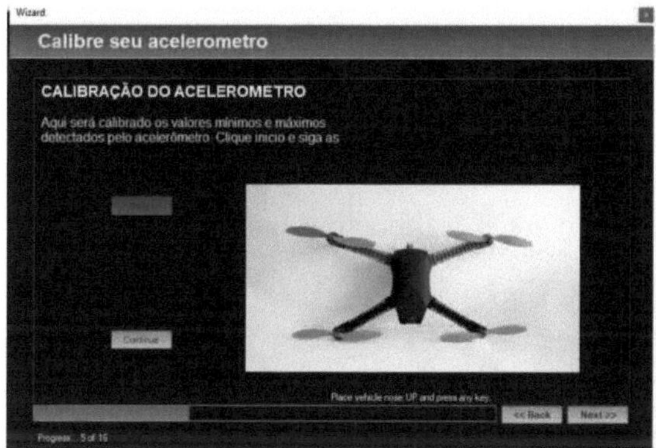

Figure 42: Accelerometer calibration Author: Prepared by the author

To finalise the calibration, place the quadcopter upside down so that the propellers touch the ground, then click "continue" and "Next" to continue with the settings.

Figura 43: Accelerometer calibration

Author: Prepared by the author

Once you've finished calibrating the accelerometer, it's also extremely important to calibrate the plate compass, also known as the bussola. It's used to indicate which position the quadcopter should follow with an autopilot, so you have to follow all the instructions on the panel. "Point all the axes to the north (left, right, front, back, top and bottom) and rotate 360 degrees on the left-right axis, for all 6 axes", by clicking on the green "calibration" button.

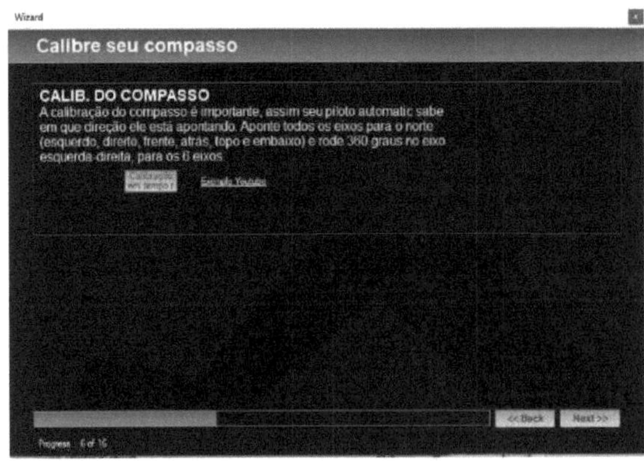

Figure 44: Calibration of the compass Author: Prepared by the author

After clicking on the "real time calibration" button, a new window will appear with a graph and a small register with the names of Compass and Samples that will change as you move the quadcopter's axis. The greater the number of Samples, the less chance there is of errors in the APM card's compass.

The graph that is used serves as a reference to record the points on the compass axis of the plate, the red dot in figure 45 is where it is

locate the quadcopter and the white dots are where you need to go during calibration, move the quadcopter in a circular fashion on all the axes, similar to calibrating a speedometer, for example; leave the quadcopter stable and turn it 360° with your hand until it reaches its starting position, doing this with all the axes until you have filled in all the blank dots.Once all the axes have been completed, the software will tell you that it has successfully captured the points. The user can also click "Next" before recording all the points, but this may cause the compass to lose data and lose information.

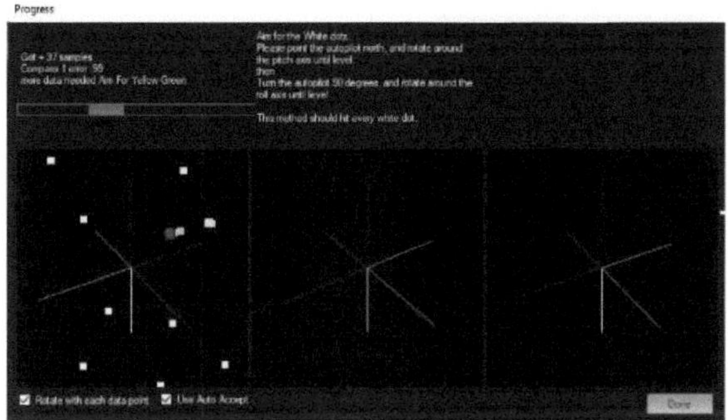

Figure 45: Calibration of the compass Author: Prepared by the author

The software will tell you that you already need to configure the axes, then click on "Nexf" to finish calibrating the compass.

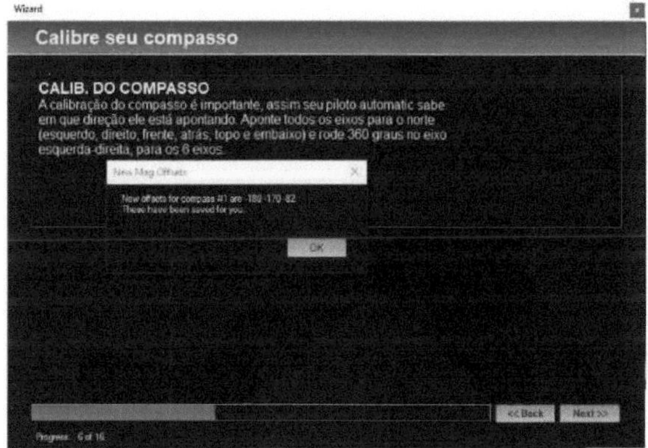

Figura 46: Calibration of the compass Author: Prepared by the author

The next step is to configure the Power Module if you have one. The configuration will take place as follows: what version of your APM card, what type of sensor and what battery capacity, fill in the three fields and press "NeX", if not just move on to the next step.

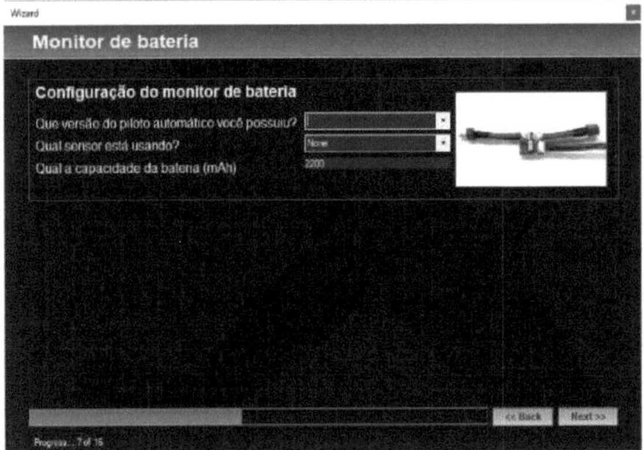

Figura 47: Battery monitor configuration Author: Prepared by the author

After configuring the Power Module we also have the option of configuring the Sonar sensor, this sensor is used to deflect obstacles during your flight, in some cases it is also used to know your height when you are very close to the ground, in this project it was not used because you do not need to use this option that the board provides, but its configuration is also very simple, just put the model that will be used and click on the enable option and continue to the next step in "Next".

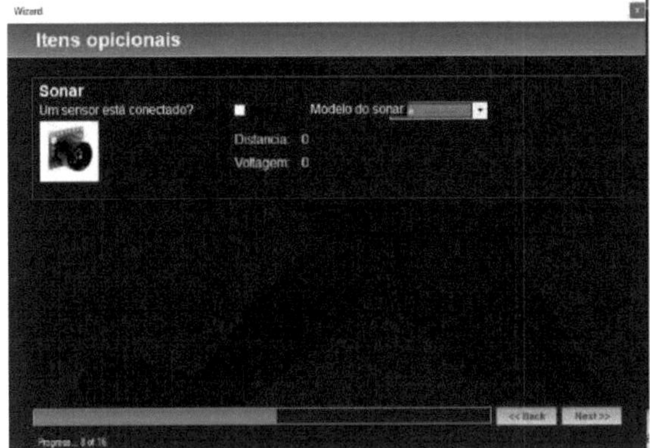

Figure 48: Sonar configuration Author: Prepared by the author

We've reached the stage of calibrating the Radio Control with the APM card. The first step is to make sure that the control is connected to the transmitter, then click on continue as shown in figure 48, which will open a new window for calibrating the radio control.

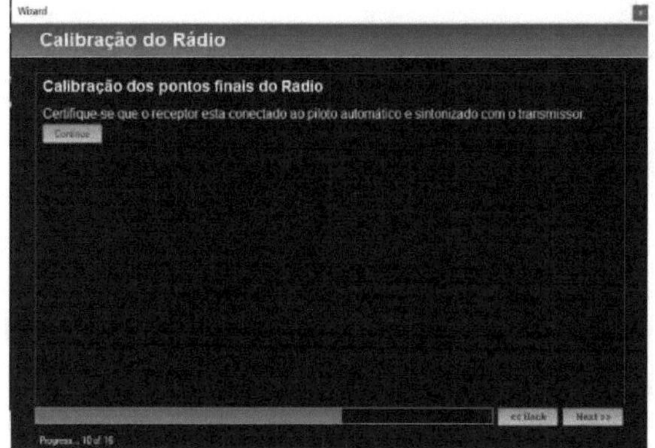

Figure 49: RC configuration Author: Prepared by the author

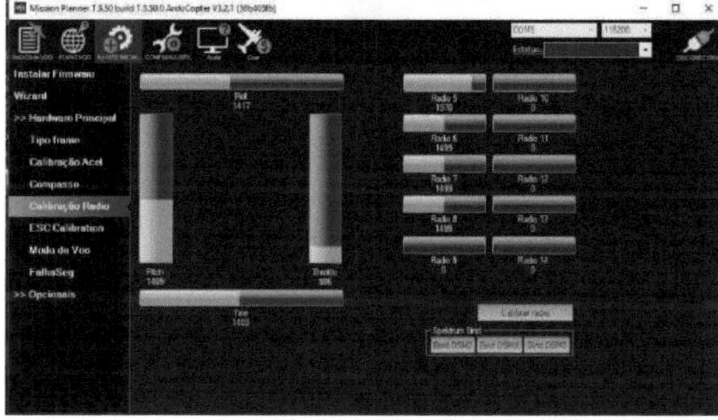

Figure 50: RC configuration Author: Prepared by the author

When you open the new image, switch on the radio control and move all the sticks to see if the green bars move, then click on calibrate radio, then a message will appear telling you to make sure that your receiver and radio are switched on correctly and that the engine has no propellers so that there are no accidents, then move all the radio sticks to mark the maximum limit and minimum limit according to the red bars in the following figure 51.

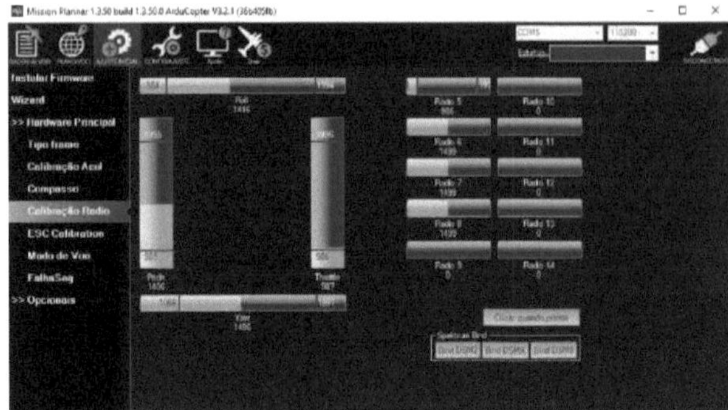

Figure 51: RC configuration Author: Prepared by the author

After doing this with all the channels connected to the APM board, click on "Click when done" and the calibration is ready with the maximum and minimum limits as shown in the next figure 52.

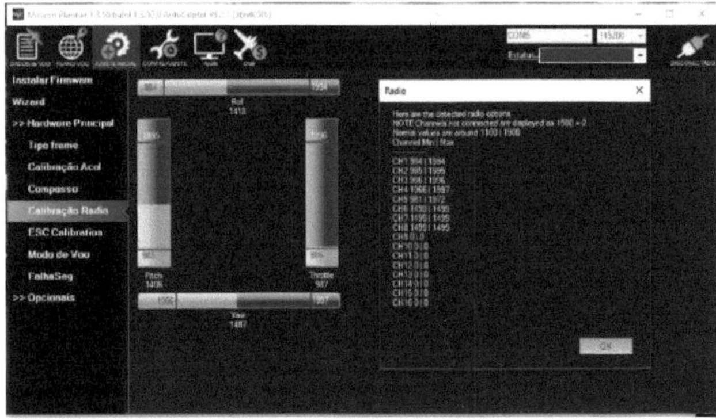

Figure 52: RC configuration Author: Prepared by the author

When the calibration is complete, click on "Next" to calibrate the flight selection mode, which has six flight mode positions as shown in figure 53.

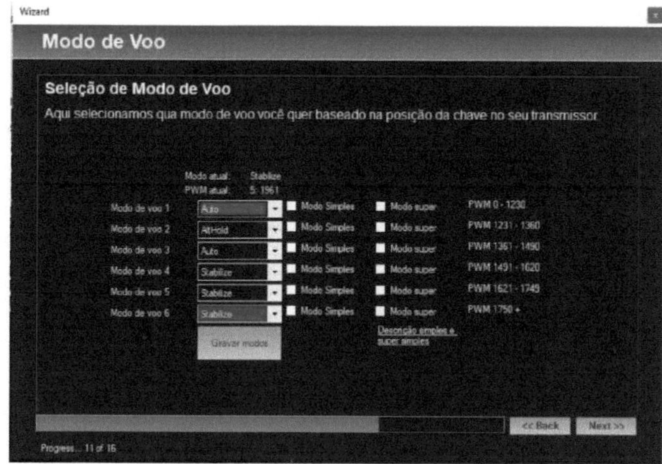

Figura 53: Flight mode configuration Author: Prepared by the author

In the case of our Radio Control we only have one position added, which is connected to channel 6 (auxiliary channel) as shown in dark green in the figure. To set the flight mode we need to know what they are for:

Stabilise - Used for flights where radio control is used, your drone will always be in the stabilised position automatically. This mode is very safe as you don't have to worry about levelling your drone, just altitude and direction, and it will stabilise itself.

Acro - A function that lets you tilt the drone according to the *sticks* on the radio control. If you do it once, it recognises it and stays that way. This function is good for filming and recording images without worrying about the angle.

Alt Hold - Similar to *Stabilise,* but maintains its altitude and removes the need to worry about height. It self-corrects with the barometer sensor or sonar sensor. This mode is recommended for those who are starting their first flights and don't need to worry about the drone's height.

Auto - Function for executing a mission via the *Mission Planner.* You can set a mission to be carried out during the flight, just set this function with the auxiliary *sticks* and it will be carried out.

Guided - A function for use with telemetry, as it works directly with the *Mission Planner* map, when you mark a position on the map it will move to it. For example, if you set a route in the GPS to collect data and return to the starting point, it will do so without you having to worry about guiding it, but be careful if there are obstacles along the way *(return-to-launch)*, it won't deviate unless it has a sonar sensor.

Loiter- This mode uses the GPS function to maintain its position, the barometer to maintain its altitude and the compass to maintain its direction. The sensors automatically keep the drone in one position without needing a command to stay in the air. It is also widely used for image recording as it is fully stabilised.

RTL) - Function to return your drone to its starting point.

Circle - A flight mode used on farms or for image recording, the drone flies in a circle with a fixed point in the middle as a reference for as long as it takes to record.

Land - If your drone has a fault or the battery is low and you can't get back to it, you can use this function to land it immediately.

Drift - A function used for racing, the drone works like an aeroplane and shoots off very quickly as instructed by the radio *sticks*, a function that is not very popular because it is easier to have accidents while flying.

Soprt - This function is similar to *Acro* but it was created specifically for filming and accessories to fix an image, the drone adjusts an angle and holds it for recording or something like that, no more than 45°.

Flip - Mode designed for acrobatics, it does a spin and returns to its position.

PosHold - It keeps the position fixed, but much smoother.

Brake - Automatically brakes the drone, once activated it doesn't accept any commands from the pilot to move forward, it automatically stops and returns to land.

Throw - Mode for launching the drone, this is done by throwing the drone upwards and it turns on and flies, but it's dangerous because you can't control the speed when it turns on.

Guided No GPS - Function for use with telemetry, as it works directly with the *Mission Planner* map, when you mark a position on the map it will move to it and land.

Follow Me - Function given to follow the pilot, you need a drone with telemetry, a notebook or mobile phone with software, a GPS or telemetry module and a camera, it recognises the pilot's position and follows him.

To choose a flight mode, click on "record". Remember to use a mode only after you understand how it works and always look for a place with open space to avoid accidents. Click on "Nexf" to continue calibrating the APM board.

The next calibration is for when battery power is low, the drone returns with telemetry and the RC signal is lost. The panel shows three modes to be calibrated as shown in image 54.

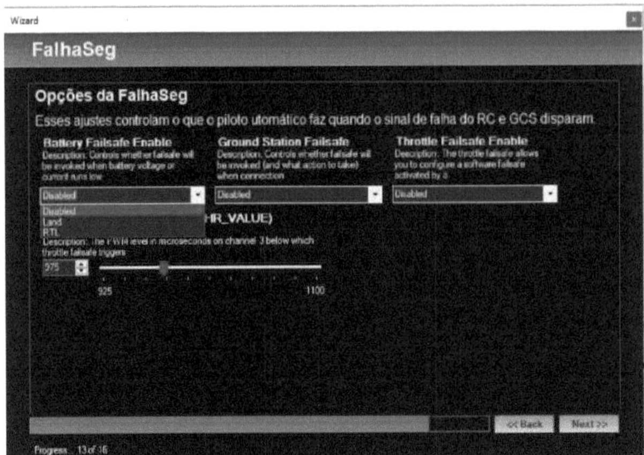

Figura 54: FalhaSeg configuration Author: Prepared by the author

The first option is given when a flight takes place and your drone runs low on battery, so you have the choice of leaving it disabled, Land (land immediately) or RTL (return home).

The second option is to use the drone guided by telemetry during a mission, and if it loses the signal you can enable it to return home or finish the mission and then return. The third configuration option, which is used when the RC signal is lost, is to return to the starting point, continue with the mission and return or land immediately.

The APM board understands when the RC loses the signal by calibrating the minimum value of the throttle (in our case it was 987), so leave the value in the bar lower than 987, as shown in image 55, which is 975.

Figure 55: RC signal loss calibration bar Author: Prepared by the author

Once you've finished, click on "Next" to calibrate *GeoFence*, which will give you the maximum distance the drone can reach, the maximum angle, maintain altitude, keep circling and so on.

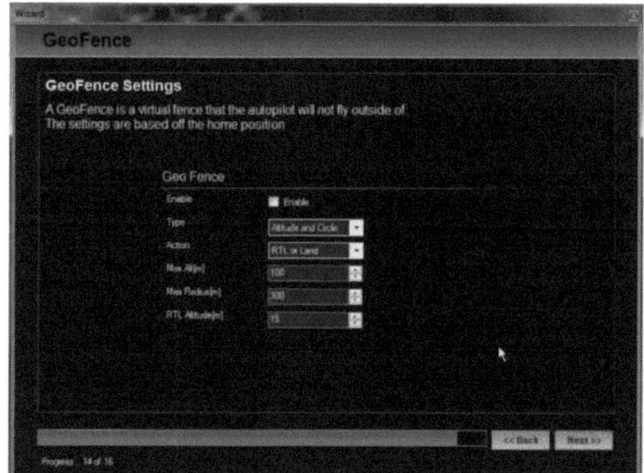

Figura 56: GeoFence configuration Author: Prepared by the author

To finish the calibration process, click on *"Uext"* which will open all the processes that have been carried out successfully.

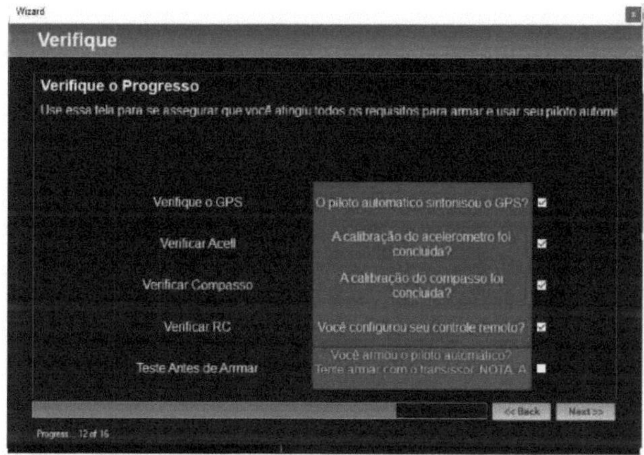

Figure 57: Finalising the settings

Author: Prepared by the author

Click on "Nexf to finalise and save your calibration.

3.5 Tests

Tests were first carried out with a board called kk2.1 - 1.5 as the first flight option. Around 6 flights were carried out with this board. The first flight took place on the Alto Araguaia campus. It lasted around 5 minutes until it lost stability and crashed on the grass. After this test, we carried out more flights in the same location, trying to adjust the KK board's configuration to achieve good stabilisation. During the last test with the board, an accident occurred with the drone, causing it to fall onto the roof and breaking one of the propellers. This board is not open source so it can be altered if necessary and it is also difficult to manoeuvre. Its configuration to make it stable is complicated, as it is necessary to distribute the weight correctly without the help of the board, it doesn't use the power of the motor to compensate for the heavier or lighter side if one of the sides has more weight. This board is used by those who already have flying experience as it is easily found on the model aeroplane market, but it didn't catch our eye to advance our project.

We did some tests with the arduino board, connecting the brushles motor, HD motor and ESC as shown in image 58, the creation of this code was to see how arduino programming works using the ESC with the motor, for future projects it is necessary to have at least a little knowledge, how the speed command is passed to the motor, this code was made through research on the internet is easily accessible if necessary.

Figure 38 Test with Arduino

Author: Prepared by the author

Below is an image of the programme developed in arduino:

```
#include <Servo.h>

Servo myservo; // Foi criado objeto Servo para controlar o servomotor

int potpin = 0; // Potpin é usado para conectar o potenciômetro
int val; // Foi criado a variável para ler o valor do pino analógico

void setup()
  {
   myservo.attach(9); // Ligue o servomotor no pino 9 ao objeto Servo
   Serial.begin(9600);
  }
void loop()
  {
   val = analogRead(potpin); // Vai ler o valor do potenciômetro (valor entre 0 e 1023)
   val = map(val, 0, 1023, 0, 179); //  Vai mapeiar o uso do servo (valor entre 0 and 180 graus)
   myservo.write(val); // Ajuste a posição do servomotor de acordo com os mapeamentes
   Serial.println(val);
   delay(15); // Aguarde um pouco até o servomotor
  }
```

Figure 59 Test with Arduino

Author: Prepared by the author

This programme made a simulation as if it were our APM controller board, the first step was to add a servo motor which

After that, we needed the input of the potentiometer chirp, which would be zero, and a variable to read how far the potentiometer would reach and pass on its value, and then we developed the code.

After carrying out tests with two boards that were not going to be used in the project, we started researching the APM board, which was then used during all the drone's flight tests. The APM board was configured and the result was as expected, so it was ready for use. Using the APM board gives the drone all the stability it needs, so at the moment there was no need to modify the code on the board, as I had planned.

The first test with the APM board was carried out as soon as it arrived, so that we could check that it wasn't faulty or had any problems. All the necessary drone accessories were installed and the programme was installed on the computer for the test, focusing on the APM board and the GPS attached to its system, all the tasks were carried out to see if the barometer, gyroscope, GPS bussola were working correctly, tests were carried out such as turning the drone completely from side to side, testing the drone's GPS in another location to see if the GPS was working correctly with the software at its coordinates on the location map.

This first test was successful even though it was not connected to the battery.

Figure 60 Test using APM Author: Prepared by the author

The second test with the drone was done with a configuration on the APM board to make a simple flight with stabilisation, we connected it to the computer and did all the configuration steps of the Software with the APM board, the execution took place without the propellers on the engine with only the software and the board. The execution of the programme with the board went perfectly without any errors being identified.

The third test was carried out using the same process, but an error appeared on the main Mission Planner screen, indicating that it couldn't find the Radio Control receiver, we changed all the batteries and checked that the Radio Control connection was correct and restarted the controller board. The software opened normally with the correct configuration, but we had to configure the RC calibration in the software again and continue with the process of checking for any other errors. The board worked perfectly, but the battery was too low to perform a flight.

The fourth test was carried out to fly the drone in the UNEMAT courtyard, the process began by letting the battery charge for 12 hours

After checking all the settings we added the propellers to the motors and tightened them with their own screws. We placed it on flat ground and removed the cable that connects the APM board to the computer and gave the signal via the radio control for it to switch on. It switched on normally and its propellers spun normally but in the weakest motor mode. When we gave the radio control more power for the drone to take off, the LED lights that are connected directly to the power supply became weak and its motors started to switch off one at a time while we watched and its ESC restarted due to lack of power, We also realised that the battery was no longer supplying the necessary power to the motors, so we switched off the controller board along with the radio control to restart the flight process, and again the same thing happened with the battery supplying little power to the motors and restarting the ESCs.

After realising that the battery was low, we put it on charge again as it could have been charged in such a way that it didn't fill up completely, but it only charged two cells when we checked the charger and it was no longer holding power for the flight, this battery was already becoming swollen, different from its normal state when it arrived, we did a more in-depth search to see what the battery's fault was and we found the answer was that its life span had been inspired and that's why it became swollen during the waiting time for the APM board to arrive, which was around 4 months.

The fifth and final test of this project was carried out with the same battery but without the propellers. We did the same process as if we were going to fly and when we started the engine it ran normally at maximum speed, requiring less energy because it didn't need to thrust the propeller and it didn't need to use as much energy from the battery.

The drone built in this project is working very well according to the tests carried out, the only reason it's not in perfect condition is that the battery has malfunctioned.

The completion of this stage of the project, with only the battery left to fly, will make it possible to continue with future projects involving Vanfs and the quadcopter itself by creating and modifying the original APM board code to develop flight routes with modifications, such as monitoring a plantation and providing all the necessary information.

At the end of the first part of this project, I'll leave you fully assembled with explanations of the configurations required for flights.

CHAPTER 4

CONCLUSION

This project involved the construction of a quadcopter. This quadcopter had all its functions approved, but there was a problem with the battery, making it impossible to fly it at the end of this project. During the testing period we got the aerodynamics to work, thus stabilising the model. To prove this stability, we used the Mission Planner programme and all the stabilisation sensors. We built a functional, self-stabilised drone that follows a route using pre-established coordinates, i.e. a flight plan that is available on the APM board itself. Once the drone was stable with the APM board, it was realised that there was no need to develop new code, as not only was there little time available, but the creation of code would only serve to personalise the drone at that point.

By using the APM flight controller board, we open the door to academics who are interested in studying automation and programming. The board has various electronic components, as well as software and hardware with components used in robotics and integrated circuits.

In addition to having all these elements in the project, in order to have a good flight with stability and precision, it was necessary to do more intensive research into the flight controller board, in which we realised that in order to modify its code and make it the way we wanted it, there wouldn't be enough time to complete this. Further studies would be needed, which would take at least another year, and only then would we be able to edit the code. Due to the short period of time we had, we were unable to make the planned changes to the code. The time we had to assemble the drone, the arrival of all the parts we had purchased and, above all, the problem with the battery, which simply stopped sending the necessary energy, meant that we were unable to complete this phase.

Another objective was to create routes, but due to unforeseen circumstances it was not possible to finalise this task. Even so, this research will serve as a step forward for the next quadcopter project, as it is fully assembled and working, and just needs the battery replaced.

If any student is interested in working with quadcopters in their dissertation, they can use the APM board code with the instructions set out on the ArduPilot board developers' website. This site has a community specifically for developers of the code used on the APM board. All you have to do is register and start practising your study, which is available to anyone who is interested. The community has several people registered, including people from different countries, so there can be an exchange of ideas and suggestions.

The importance of this quadcopter is due to the fact that it is an innovative project on this university campus. This is the first time the campus has presented final-year coursework using aeromodelling technology. We hope for the same success in future tests and trials with students who want to work with quadcopters, in terms of using these results, as well as systematically correcting and improving them in the next stages, including studying and programming the flight controller board.

REFERENCES

ALBERTO, E, M, F. **Low-cost GPS/INS inertial navigation system with error compensation by artificial neural networks**. Thesis (Doctorate) - National Institute for Space Research, São José dos Campos, 2011. 150 p.

ANAC. **Drone Registration and Certification.** Last updated on 02/05/2017. Available at: <http://www.anac.gov.br/assuntos/paginas-tematicas/drones/registros-e-certificados-de-drones>. Accessed on: 24/08/2017.

ANAC. **General Requirements for Civil Unmanned Aircraft.** P. 5-6 RBAC-E n°94 with resolution n°419 of 2 May 2017. Available at: <http://www.anac.gov.br/assuntos/legislacao/legislacao-1/rbha-e-rbac/rbac/rbac-e-94-emd-00>. Accessed on 24/08/2017.

ANDRADE, R, D, N. **Creation of a Test Station for an Autonomous Micro-Helicopter.** Master's Dissertation - University of Madeira, Master's Degree in Computer Engineering, 2012. Available at: <http://digituma.uma.pt/bitstream/10400.13/595/1/MestradoRa%C3%BAIAndrade.pdf> Accessed on: 30/11/2017.

BARATO, B. **Design of a control system for unmanned aerial vehicles**. University of

São Paulo São Carlos Engineering, 2014.

BRANDÃO, M, P. et al. UAV Activities in Brazil. First Latin-American Uav Conference. Panama, 2007.

CARMO, B. R. **Development of a fully configured UAV for Precision Agriculture applications in Brazil** - CNPDIA- EMBRAPA, São Carlos, São Paulo, 2011.

DEMOLINARI, C, H. **Projeto de Construção um Drone Hexacóptero** - Monografia, Mechanical Engineering Course, Universidade Federal Fluminense, Niterói, 2016.

FILHO, G. L. S.; RUDIGER, G. T.; NASCIMENTO, J. P. M. **Quadricopter**. 2011. vi,33 f. Monografia-Pontifica Universidade Tecnológica Federal do Paraná, 2011. Disponívelem : <http://paginapessoal.utfpr.edu.br/msergio/portuguese/ensino-de-

fisica/oficina-de-integracao-ii/oficina-de-integracao-ii/Monog-11-2-Quadricoptero.pdf>. Accessed on: 29/10/2015.

FOUR, R.; RAMOUTAR, E.; ROMEO, J.; COPELAND, B. **Operational Characteristics of Brushless DC Motors** - University of the West Indies, St.Augustine, Trinidad. 2007;

HARDGRAVE, L. **The pioneers. Aviations and Aeromodelling** 2005. Available at :<http:/www.ctie.monash.edu.au/hargrave/denny.htm>. Accessed on 20/01/2017.

HONÓRIO, L. M.; OLIVEIRA, E. J.; ALVES, A. S. C. **Study and Application of Embedded Control Techniques for Quadcopter Flight Stabilisation**. 2012. vi,121 f. Thesis (Electrical Engineering) - Federal University of Juiz de Fora, 2012. Available at: <http://blog.opovo.com.br/asaseflaps/quadricopteros-aprenda-um-pouco- more-about-this-wonderful-machine/> Accessed on: 28/10/2015.

JUNIOR, A. A. P.; SILVA, C. M. G.; SANTOS, J.; TEIXAIRA, L. J. **Quadricopter**, Rio de Janeiro, 2013. p. 5, 6, 7. Available at: <http://www.crea-rj.org.br/premiocrearjniemeyer/files/2014/10/ETSS-Eletrot%C3%A9cnica.pdf> Accessed on: 10/11/2015.

KEANE, J. F., CARR, S. S., "A Brief History of Early Unmanned Aircraft", Johns Hopkins APL Technical Digest, V32, N3, 2013

LONGHITANO, G, A. **UAVs for remote sensing: applicability in the assessment and monitoring of environmental impacts caused by accidents with dangerous cargo.** Dissertation (Master's in Transport Engineering) - Polytechnic School of the

University of São Paulo, 2011.

MEDEIROS, F.A. **Development of an unmanned aerial vehicle for application in precision agriculture**, 2007. 102f. Dissertation (Master's in Agricultural Engineering) - Federal University of Santa Maria, Santa Maria.

MILESKI, A, M. **A history of high technology.** In: Revista Tecnologia e Defesa, a.20, n.92, p. 42-61, 2007.

OLIVEIRA, D. Unmanned Aircraft - ANT - History at CTA and Perspectives. **Aeronautical Systems Division -ASA**, São José dos Campos: CTA, 2005.

RODRIGUES, H. K. **Design of an Efficient Autonomous Quadcopter**.2014 vi.42 f. Final Report. Polytechnic School of the University of São Paulo, 2014. Available at: < http://www.usp.br/ldsv/wp-content/uploads/2014/10/Relat%C3%B3rio_FINAL_IC_AEP_HENRIQUE_KO KRON_RODRIGUES.pdf> Accessed on 19/10/2015.

ROMERO, E, L; POZO, F, D; ROSALE, J, A. **Quadcopter stabilisation by using PID controllers** - Universidad de Las Américas- Ecuador, 2014.

ROSKAM, J. Airplane Flight Dynamics and Automatic Flight Controls, DARcoporation, Lawrence, 2003.

SANTOS, A. Development of a Self-Stabilising and Self-Guided Quadcopter - State University of Mato Grosso - UNEMAT, 2016

SANTOS, J.; TEXEIRA, L.G. Quadricopter. - Monograph, Sandra Silva Technical School 2013.

SANTOS, A. B.; SANTOS, B. J.; LISBOA, L. Drones, models, uses and applications.2014. Available at :< https://dlq8vi77lxj74.cloudfront.net/media/e55e91b2cc22ba117ba8d1546537f7a4c037cf67/d675ef0f9d8a788f6d5a3803769697a930d4d5d0/1411838094/dronemodelosusoseaplicac oes.pdf> Acesso em 19/11/2015.

SILVA, L, K.; MORAIS, S, A.; MORAIS, S, J.; GEDRAITE R. Hardware for advanced control of a Quadcopter unmanned aerial vehicle - Universidade federal de Uberlândia- minas Gerais 2013.

SILVEIRA, V. Brazil is already developing an unmanned vehicle. Available at:<http://www.defesanet.com.br/tecno/vant/>. Accessed on: 09/01/2017.

STUDART, A. Quadricopter, learn a little more about this marvellous machine. 2015.Available at: <http://blog.opovo.com.br/asaseflaps/quadricopteros- learn-a-little-more-about-this-marvellous-machine/> Accessed on 29/10/2015.

VETTORAZZI, C.A.; ANGULO FILHO, R.; COUTO, H.T.Z. Global Positioning System - GPS. Engenharia Rural, Piracicaba, v.5, n.2, p.61-70, 1994. Available at:

<http://www.hobbyking.com/hobbyking/store/__43709__Afro_ESC_20Amp_Multi_rotor_Motor_Speed_Controller_SimonK_Firmware_.html> Accessed on 30/08/2016.

VIEIRA, J, C, S. QuadRotor Aerial Mobile Platform, Monograph, University of Minho, 2011.

ZACCARELLE, L, C. The Electronic Speed Control ESC. 2012. Available at: <http://aeromodelismoiniciante.blogspot.com.br/2011/01/o-eletronic-speed-control- esc.html> Accessed on 01/09/2016.

Printed by Books on Demand GmbH, Norderstedt / Germany